W9-CID-364

CONVOLUTION INTEGRAL EQUATIONS

Convolution Integral Equations

with special function kernels

H. M. SRIVASTAVA

Professor of Mathematics
University of Victoria, British Columbia, Canada

R. G. BUSCHMAN

Professor of Mathematics
University of Guelph, Ontario, Canada

A HALSTED PRESS BOOK

JOHN WILEY & SONS

New York London Sydney Toronto

Copyright ©1977, WILEY EASTERN LIMITED

Published in the U.S.A., Canada, Latin America
and the Middle East by Halsted Press,
A division of John Wiley & Sons, Inc., New York

Library of Congress Cataloging in Publication Data

Srivastava, H M
    Convolution integral equations, with special function
kernels.

  "A Halsted Press book."
  Bibliography : p.
  Includes indexes.
    1. Volterra equations—Numerical solutions.
2. Convolutions (Mathematics). 3. Kernel functions.
I. Buschman, R.G., joint author. II. Title
QA431.S64   515´.45   76-52979
ISBN 0-470-99050-3

Printed in India at Rekha Printers Private Ltd.,
New Delhi 110020.

To Rusty, who types

CONTENTS

CHAPTER 3

METHODS AND EXAMPLES

CHAPTER 4

MISCELLANEOUS RESULTS

PREFACE

A Volterra integral equation of the first kind can be written
in the form

$$(0.1) \qquad \int_{a}^{x} k(x,t)\, f(t)\, dt = g(x),$$

where $a > -\infty$. It is known that for certain cases this equation
can be converted into a Volterra equation of the second kind by
differentiation. It is our purpose to consider alternative methods
of solving the equations of the first kind where the kernels are
given special functions, since quite often the requirements for
conversion to equations of the second kind are not satisfied. In
particular, we concentrate on a survey of the kernels and of the
methods for which explicit integral (or differential) inversion
formulas can be obtained.

In Chapter 1 we give a discussion of those kernels which are
available in the literature with references to the lengthy bibliography
which has been collected. Chapter 2 consists of a discussion of
relations between various forms of the equations, a discussion of
the question of uniqueness of solutions, taking into consideration
the Theorem of Titchmarsh and the work of Gesztelyi, and a discussion
of the related integral transformations and fractional integrals.
A description of each of several methods is given in Chapter 3 with
an explicit, simple example for the illustration of each method.
Further, in Chapter 4 we include some miscellaneous results and mention
various problems which may be worthy of further investigation.

Some notations which we use for special functions appear in the Appendix, and, in general, these notations are the ones used in "Higher Transcendental Functions" by A. Erdélyi et al. [41]. We shall refer often to "Tables of Integral Transforms", by A. Erdélyi et al. [42] for integrals. In both of these cases we use abbreviations such as [41:2.1(2)] to refer to the reference numbered 41, chapter 2, section 1, formula (2).

Integral transformations are mentioned as methods of solving integral equations of the first kind in a number of books; for example, R.V. Churchill [27], V.A. Ditkin and A.P. Prudnikov [30], Gustav Doetsch [32,34], Balth. van der Pol and H. Bremmer [80], Ian N. Sneddon [115, 117], and E.C. Titchmarsh [140]; the operational calculus of Mikusiński is used by Lothar Berg [1] and Arthur Erdélyi [37]. For discussions of convolutions and their properties, besides the already listed texts, one can refer to Gustav Doetsch [33] for a lengthy discussion of the simple convolution and to István Fenyö [44] and to E. Gesztelyi [48] for generalizations. J.G. Mikusiński and Cz. Ryll-Nardzewski [72] give a table relating the class to which the convolution product belongs to the classes from which the factors come.

For lack of a brief common terminology to describe the method in which a key integral involving two kernels k and k_1 is computed and then a direct verification of the proposed solution is made by substitution into the original equation, we introduce the expression "resolvent kernel method". Further discussion and an example of this method is contained in section 2 of Chapter 3.

Although we have restricted our discussion mostly to forms related to equation (0.1), we do include a few cases of somewhat related equations of the first kind for which the interval is of the form (x,∞); the cases of other equations of the first kind where the intervals for integration are (a,b) or $(-\infty, +\infty)$ are quite different problems. Since a large variety of physical problems lead to convolution integral equations with special function kernels, it is hoped that the types of solutions explored here will be useful in the disciplines of applied mathematics, theoretical and quantum mechanics, and mathematical physics.

The collaboration on this project was initiated, and a preliminary draft was prepared, during the academic year 1972-73 while the second author was at the University of Victoria on sabbatical leave from the University of Wyoming. This monograph has since been revised and updated a number of times.

Victoria
and
Guelph
June 1976

H. M. Srivastava

R. G. Buschman

CHAPTER 1

LITERATURE ON SPECIAL FUNCTION KERNELS

1.1 Algebraic kernels

The books mentioned in the introduction as well as many of the
papers which will be mentioned in this chapter treat Abel's equation,

$$(1.1.1) \qquad \int_a^x (x-t)^{-\alpha} f(t) \, dt = g(x), \qquad\qquad 0 < \alpha < 1,$$

as an example in view of its simplicity and also because it follows
as a special case of other kernels. This equation has a long history
which we have not specifically pursued. The inversions of certain
fractional integrals are definitely related to the problem of the
solution of Abel's equation or its generalizations; see, for example,
L.S. Bosanquet [10], A. Erdélyi [35, 36], H. Kober [63], J.S. Lowndes
[68], and C.V.L. Smith [114]. Both intervals (a, x) and (x, ∞)
have been considered in the literature.

Certain other equations can be reduced to Abel's equation. For
example, the equation

$$(1.1.2) \qquad \int_a^x \bigl(k(x) - k(t)\bigr)^{-\alpha} f(t) \, dt = g(x), \qquad 0 < \alpha < 1,$$

can sometimes be reduced to an Abel equation by a simple change of
variables; see J. Burlak [13], A.N. Hovanskii [56], R.P. Srivastav
[121], and E. Gesztelyi [48].

The related problem involving the Hadamard finite part of a divergent integral when $\alpha \geq 1$ is considered by T.K. Boehme [8], P.L. Butzer [21, 22], Erdélyi [37], Fritz Rühs [90], and Klaus Wiener [145, 146] and a series of related papers.

For integral transformation and operational calculus techniques for algebraic kernels see also Butzer [20], Ky Fan [43], C. Fox [46], Ram P. Kanwal [61], and Sneddon [116].

Although we are not considering numerical methods here, a recent paper by Richard Weiss [142] indicates references in that direction.

1.2 Exponential, logarithmic, and trigonometric kernels

Some kernels involving the exponential function, for which the technique of Laplace transforms is applied, have been treated by Kanwal [61] and by D.O. Reudink [88]. The general problem of finding inversion integrals which involve the same kernel function was studied by Reudink [loc. cit.]; Fenyö [45] also treated this problem of Reudink by means of Mikusiński operators. Kernels which involve logarithms were discussed by S. Colombo [29], Ky Fan [43], and L. Poli [81], and for Hadamard finite parts by Wiener [145]. Trigonometric and hyperbolic functions as kernels are included in the works of Butzer [20], Erdélyi [37], Kanwal [61], van der Pol and Bremmer [80], Reudink [88], K.C. Rusia [98], and Sneddon [115, 117]. Generalized hyperbolic and trigonometric functions [41:18.2] were used as kernels by P.L. Bharatiya [4].

1.3 Chebyshev polynomials

Much of the recent interest in these equations seems to have
been initiated by the papers of Ta Li [64, 65] who solved

$$(1.3.1) \qquad \int_x^1 \left(t^2 - x^2\right)^{-1/2} T_n(t/x) \, f(t) \, dt = g(x),$$

in which $T_n(x)$ denotes a Chebyshev polynomial of the first kind of
degree n in x. This equation arose from problems in aerodynamics.
He used the resolvent kernel method and he needed to evaluate the
key integral involving the product of kernels,

$$(1.3.2) \qquad \int_x^v \left(t^2 - x^2\right)^{-1/2} T_n(t/x) \left(v^2 - t^2\right)^{-1/2} T_{n-1}(t/v) \, dt.$$

In part, the effort of further research has been directed toward
finding simpler methods of evaluating such integrals when they are
not already tabulated, as well as toward finding simpler methods of
discovering solution forms involving other kernels. Other efforts
have been made in obtaining alternative methods for solving the
equation; Sneddon [116, 117] has used the Mellin transformation and
D.V. Widder [144] used the Laplace transformation after making a change
of variables. Dietrich Suschowk [138] considered a related equation
using a formula for $T_n(x)$, manipulations, and a limiting process.

In the case of the Chebyshev kernels, as well as in a number of
other cases which follow, the solution could be obtained by specializing
the solution of an integral equation with a more general function as
the kernel.

1.4 Legendre polynomials and functions

. In a number of papers kernels which involve Legendre polynomials
have been considered; for example, the simple equation

$$(1.4.1) \qquad \int_x^1 P_n(t/x) \ f(t) \ dt = g(x),$$

was solved by Buschman [14] using the resolvent kernel method, by
Erdélyi [38] using Rodrigues' formula followed by successive integrations
and differentiations, by Sneddon [116] using the Mellin transformation,
and by Widder [144] using the Laplace transformation. A similar
equation with a kernel involving generalized Legendre polynomials
which are closely related to special cases of Jacobi polynomials was
treated by R.P. Singh [112]. The equation

$$(1.4.2) \qquad \int_0^x P_n\bigl(\cos \alpha(x-t)\bigr) \ f(t) \ dt = g(x),$$

and its solution appeared in the list of van der Pol and Bremmer [80]
for $\alpha = 1$ and is also considered by B.R. Bhonsle [6]. Mikiharu
Terada [139] applied Mikusiński operators to a similar equation with
kernel $P_n\bigl(e^u\bigr)$.

A quite different method using partial differential equations
was invoked by A.G. Mackie [69] to solve the equations

$$(1.4.3) \qquad \int_0^x P_n(t/x) \ f(t) \ dt = g(x)$$

and

$$(1.4.4) \qquad \int_0^x P_n(x/t) \; f(t) \; dt = g(x).$$

The first of these equations was also treated in the series of three papers, R.A. Sack [100], Ll.C. Chambers and R.A. Sack [25], and Ll.C. Chambers [24]. Equations of this type have quite different properties from those mentioned in the preceding paragraph; the solutions need not be unique because of orthogonality relations. This is discussed in Chapter 2.

For equations with kernels involving Legendre functions; specifically

$$(1.4.5) \qquad \int_x^1 \left(t^2 - x^2\right)^{\lambda/2} P_\nu^\lambda(t/x) \; f(t) \; dt = g(x)$$

and

$$(1.4.6) \qquad \int_1^x \left(x^2 - t^2\right)^{\lambda/2} \underset{\sim}{P}_\nu^\lambda(t/x) \; f(t) \; dt = g(x),$$

with $\mathrm{Re}(\lambda) < 1$, solutions were given by Buschman [16] by means of the resolvent kernel method. Erdélyi [39, 40] used the method of fractional integration to factor the operator into a product of two fractional integrals and hence obtained the solution for the range of integration (a, x); further, in the second paper, he treated the problem of Hadamard finite parts. The Mellin transform was applied by Sneddon [116] to solve the equation for the range (x, ∞) with x > 0. Solutions can be obtained as special cases of $_2F_1$-kernels such as has been done recently by Tilak Raj Prabhakar [86].

1.5 Gegenbauer and Jacobi polynomials

Solutions to equations of the form

$$(1.5.1) \qquad \int_x^1 \left(t^2 - x^2\right)^{\lambda - 1/2} C_n^\lambda(t/x) \, f(t) \, dt = g(x),$$

which involve Gegenbauer polynomials in the kernel, were given by

Buschman [15] for a special case, by Theodore P. Higgins [53], and

by K.N. Srivastava [129] (where the corrected results of [127] appear);

the resolvent kernel method was used. Sneddon [116] used the Mellin

transformation and Prabhakar [86] listed the solution among special

cases of the $_2F_1$-kernel.

In an earlier paper K.N. Srivastava [126] applied the Hankel

transformation to obtain solutions to similar equations where the

ranges of integration are $(0, x)$ and (x, ∞).

Equations involving the Jacobi polynomials in the form

$$(1.5.2) \qquad \int_x^1 \left(t^2 - x^2\right)^\alpha P_n^{(\alpha, \beta)}\left(2t^2/x^2 - 1\right) f(t) \, dt = g(x),$$

have been studied by K.C. Rusia [91] and K.N. Srivastava [128, 130,

132, 133], both using resolvent kernels. On the other hand, Rusia [95]

used Rodrigues' formula for the case in which α is a non-negative

integer, and in [97] he applied the Hankel transformation. Prabhakar

[86] listed the equation among the special cases of $_2F_1$-kernels.

C. Singh [110] used the fractional derivative form of Rodrigues'

formula to solve the equations with the kernels involving

$P_n^{(\alpha, \beta)}(1 - 2x/t)$; Bhonsle [7] considered kernels involving $P_n^{(\alpha, \beta)}\left(1 - 2x^2/t^2\right)$.

Jacobi functions occurred in the kernels of K.N. Srivastava [134] where the ranges of integration (0, x) and (x, ∞) were considered and the Hankel transformation applied.

1.6 Laguerre and Hermite polynomials

In these cases the equation is taken in the form of a Laplace convolution. For the Laguerre polynomial kernels we have the equation

$$(1.6.1) \qquad \int_0^x e^{b(x-t)} (x-t)^\alpha L_n^{(\alpha)}(x-t) \, f(t) \, dt = g(x),$$

which was treated by Rusia [92]. For the case in which α is a non-negative integer Rusia [95] used Rodrigues' formula. For the case b = 0 Buschman [17] used Mikusiński operators, P.R. Khandekar [62] used the Laplace transformation, and K.N. Srivastava [135] used the resolvent kernel method. For $\alpha = b = 0$ Hisachi Choda and Marie Echigo [26] used Mikusiński operators and van der Pol and Bremmer [80] and Widder [143] used the Laplace transformation.

M.T. Shah [114] applied the resolvent kernel method and Laplace transforms to the equation

$$(1.6.2) \qquad \int_0^x (x-t)^{-1/2} H_{2n}\left(\alpha(x-t)^{1/2}\right) f(t) \, dt = g(x),$$

which involves Hermite polynomials. Notice that since

$$(1.6.3) \qquad H_{2n}(x) = (-1)^n \, 2^{2n} n! \, L_n^{(-1/2)}(x^2),$$

Shah's solution of (1.6.2) is evidently contained in that of the well-treated equation (1.6.1) with $\alpha = -1/2$ and b = 0.

Prabhakar [82] obtained Laguerre and Hermite kernels as special cases of his results on confluent hypergeometric functions.

1.7 Bessel functions

First we consider the equation

$$(1.7.1) \qquad \int_0^x \left(x^2-t^2\right)^{\nu/2} J_\nu\left(c(x^2-t^2)^{1/2}\right) f(t) \, dt = g(x),$$

which involves the Bessel function of the first kind. This and the associated equation on (x, ∞) have been solved by Burlak [12] by using the Laplace transformation method, by Srivastav [122] by the resolvent kernel method, related equations by J.S. Lowndes [68] by the fractional integral method, and L^2 solutions have been investigated by Kusum Soni [119, 120] by the fractional integral method. S.L. Kalla [59, 60] investigated related equations which contain an exponential factor in the kernel on the ranges (a, x) and (x, b).

Solutions for the case of zero order Bessel kernel have been given by Bharatiya [3], J. Heinhold [52], Maurice Parodi [77, 78], van der Pol and Bremmer [80] and Sneddon [117] using the Laplace transformation; by Butzer [20] and by Erdélyi [37] using Mikusiński operators; and for the related case with an exponential factor by Reudink [88]. Rühs [90], on the other hand, considered the Hadamard finite part case associated with the kernel $(t-x)^{-1} J_0(t-x)$.

Soni [118] used fractional integration in her investigation of L^2 solutions of

$$(1.7.2) \qquad \int_0^x J_0\left(t(x^2-t^2)^{1/2}\right)(xt)^{1/2} f(t) \, dt = g(x).$$

Methods involving partial differential equations were used by Mackie

[69] to solve a similar equation with kernel involving I_0. The Laplace transformation was applied by J. Heinholt [52] to treat the case of the kernel

$$(1.7.3) \qquad k(x,t) = \int_t^x J_0\left(\alpha(x-u)\right) J_0\left(\alpha(u^2-t^2)^{1/2}\right) du,$$

associated with an equation of the form (0.1). Rusia [93] considered the equation (1.7.1) with I_n replacing J_ν and Prabhakar [82] obtained a kernel involving I_ν as a special case of the $_1F_1$-kernel.

C. Singh [107] worked with Kelvin's function, ber(ax) as a kernel and used the Laplace transformation, and in [106,108] he used the K- and H-transform inversion formulas along with the Mellin transformation to obtain solutions to equations involving the kernels $K_\nu\left(z(xy)^{1/2}\right)$ and $H_{m,\nu}\left(z(xy)^{1/2}\right)$ (Struve's function) over both the ranges $(0, x)$ and (x, ∞).

1.8 Kummer's and Whittaker's functions

For equations of the form

$$(1.8.1) \qquad \frac{1}{\Gamma(c)} \int_0^x (x-t)^{c-1} \, _1F_1\left(a;c;\lambda(x-t)\right) f(t) \, dt = g(x),$$

which involve the confluent hypergeometric function $_1F_1$, Mikusiński operators were applied by Buschman [19], fractional integration by G. Mustafa Habibullah [51] and Prabhakar [82] who also considered the ranges (a, x) and (x, b) in [85], the resolvent kernel method by Rusia [96] and K.N. Srivastava [131], and the Laplace transformation by Sneddon [117] and Jet Wimp [147]. A similar equation with an

exponential factor in the kernel was treated by Rusia [94]. C. Singh
[105,109] and K.N. Srivastava [136] considered the kernel expressed
in terms of Whittaker's function $M_{k,m}(z)$, and used the resolvent
kernel method.

Solutions for the special cases of $_1F_1$ which involve incomplete
Γ-functions were given by Bhonsle [7], R.N. Jagetya [57], Reudink
[88], and Rusia [98]; Reudink and Rusia also listed an error function
kernel. Bateman's function kernels were considered by Bharatiya [2]
and Rusia [93]. Habibullah [50] considered kernels involving Shively's
polynomials.

1.9 Gauss' hypergeometric function

E.R. Love [66, 67] has given a thorough discussion of kernels
involving the Gauss function by using fractional integrals and thereby
obtaining factorization of the operators for the equations of the form

$$(1.9.1) \qquad \frac{1}{\Gamma(c)} \int_0^x (x-t)^{c-1} \, _2F_1(a,b;c;1-x/t) \, f(t) \, dt = g(x),$$

$$(1.9.2) \qquad \frac{1}{\Gamma(c)} \int_0^x (x-t)^{c-1} \, _2F_1(a,b;c;1-t/x) \, f(t) \, dt = g(x),$$

$$(1.9.3) \qquad \frac{1}{\Gamma(c)} \int_x^\beta (t-x)^{c-1} \, _2F_1(a,b;c;1-x/t) \, f(t) \, dt = g(x),$$

$$(1.9.4) \qquad \frac{1}{\Gamma(c)} \int_x^\beta (t-x)^{c-1} \, _2F_1(a,b;c;1-t/x) \, f(t) \, dt = g(x),$$

where $\beta \overset{<}{=} \infty$.

A number of other recent papers consider similar equations. Buschman [18] used fractional integrals on (1.9.2), Higgins [54] used fractional integrals on (1.9.4) for $\beta = 1$, van der Pol and Bremmer [80] used the Laplace transformation on (1.9.2), Prabhakar [86] considered equations similar to (1.9.3) and (1.9.4) with $\beta < \infty$ by use of fractional integrals, Sneddon [116] used the Mellin transformation on (1.9.3) with $\beta = 1$, and Wimp [147] used the Laplace transformation on (1.9.3) with $\beta = 1$. G. Mustafa [75] used fractional integrals on a special case of an equation related to (1.9.1).

Many special cases of interest can be obtained from Love's results by specializing the parameters of $_2F_1$ and by restricting the domains of non-zero values of the functions f and g. A number of these cases have been computed and are listed in the Inversion Tables.

1.10 Generalized hypergeometric and other functions

Among the generalized hypergeometric functions those which have been used in kernels are $_1F_2$ by Reudink [88], Appell's function F_3 by Higgins [53], two cases of Φ_1 by Prabhakar [87], the confluent hypergeometric function of r variables Φ_2^r by H.M. Srivastava [124], and Φ_3 by K.S. Sevaria [103]. O.I. Maricev [70] has considered equations involving the functions Ξ_2 and F_3, which arose from the study of the Cauchy and the Tricomi problems for a certain hyperbolic partial differential equation; solutions are obtained in terms of fractional integrals. R.N. Jagetya [57] considered kernels involving infinite series of $_2F_2$ and of $_2F_3$, but the reduction to a single term gives $_1F_2$ and $_1F_1$ kernels. A few additional special cases of $_0F_2$, $_1F_3$, and Ψ_1 have been computed from the tables of Laplace transformations and are entered in the Inversion Tables.

The elliptic functions considered are θ_2 by Rusia [93], θ_3 by van der Pol and Bremmer [80] and θ_3' by Bhonsle [7]. Mittag-Leffler functions were used as kernels by Bharatiya [5] and a generalization by Prabhakar [83, 84]. T.N. Srivastava and Y.P. Singh [137] used Wright's generalized Bessel function, J_ν^μ.

Ben C. Johnson [58] used the resolvent kernel method to study equations for which the kernels can be represented as Mellin-Barnes integrals. His kernel functions ξ, $\bar{\xi}$, and ξ^* are the generalized hypergeometric functions of N.E. Nørlund [76] which are special cases of the G-function and his functions R and T are special cases of the H-function which generalize those functions studied by Boersma [9]. H.M. Srivastava [123] showed how the systematic use of the theory of the Mellin transformation leads to a simple procedure by means of which a certain class of integral equations may be solved. Indeed, he applied this technique, which presupposes the existence of the Euler transform of the kernel as well as the Mellin transform of the Euler transform, to solve the integral equation

$$(1.10.1) \qquad \int_y^\infty (x-y)^{-\alpha} \; H_{p,q}^{m,n}[x-y] \; f(x) \; dx = g(y), \qquad y \overset{\geq}{=} 0,$$

where $H_{p,q}^{m,n}[\zeta]$ denotes the H-function of Fox [47]. Further, the equation

$$(1.10.2) \qquad \int_0^x (x-t)^{\rho-1} \, H_{p,q}^{m,n}[x-t] \, f(t) \, dt = g(x)$$

has recently been studied by H.M. Srivastava and R.G. Buschman [125], see Example X in section 4.2.

Some other related general ideas are worthy of mention here. In the review of [2], it was pointed out by N.K. Chakravarty [23] that the Laplace transform method applies well to all equations of the type

$$(1.10.3) \qquad \int_0^x k(x-t) \, f(t) \, dt = g(x), \qquad 0 \leqq x < \infty,$$

for which the kernel has a Laplace transform of the form

$$(1.10.4) \qquad K(p) = p(p-\alpha)^m (p-\beta)^{-n}.$$

It should be noted that the Mellin transformation method, or the related fractional integral method, is especially suitable for attempts at solving equations of the form

$$(1.10.5) \qquad \int_a^x k(x/t) \, f(t) \, dt = g(x),$$

or for the range of integration (x, b), if the kernel has a Mellin transform involving Γ-functions. In the discussion of the problem of Reudink, I. Fenyö [45] considered kernels of the form of convolution product involving k factors like

$$(1.10.6) \qquad \frac{t^{m_j/2-1} \, e^{\lambda_j t}}{\Gamma(m_j/2)}.$$

The possibility of the reduction of kernels, which involve generalized hypergeometric functions, by use of fractional integration was mentioned by Buschman [18]. On the other hand, L.A. Sahnovich [101] has considered the problem of finding the nth root of the operator K, where

$$(1.10.7) \qquad K f(x) = \int_0^x k(x-t) f(t) \, dt,$$

and of the reduction of K to the simplest form. It should also be mentioned that integral transform techniques often lead to the solution of related equations by examination of the resulting integral form of the solution and the consideration of this form as itself an equation.

CHAPTER 2

SOME BASIC PROPERTIES

2.1 Convolutions and relations among equations

We shall refer to the convolution defined by

$$(2.1.1) \qquad (k * f)(x) = \int_0^x k(x-t) \, f(t) \, dt,$$

as the "Laplace convolution" because of its natural connection with
the Laplace transformation. A development of the properties of this
convolution is given in Doetsch [33:2,14-15]. An investigation of
the relation between the class of functions to which the convolution
belongs and the classes to which the factors belong is given by J.G.
Mikusiński and Cz. Ryll-Nardzewski [72]; the results are tabulated
on page 57 of that paper. Discussions of the convolution ring appear,
for example, in the works of Berg [1], Mikusiński [71], and Erdélyi
[37].

In his work on operational calculus Gesztelyi [48] proved the
following result involving generalized convolutions.

<u>Theorem</u>. Let $K(x,t)$ be strictly monotone in the variable t
for every fixed x and continuous in

$$(2.1.2) \qquad D = \{(x,y) \, | \, a \overset{\le}{=} t \overset{\le}{=} x < \infty\},$$

and let $\phi(x)$ be normalized by

$$(2.1.3) \qquad \phi(a) = 0, \quad \phi(x) = \frac{1}{2}\{\phi(x+) + \phi(x-)\} \, ,$$

and let it be a monotone function in $a \leq x < \infty$. If $C(a,\infty)$ has no divisors of zero with respect to the generalized convolution

$$(2.1.4) \qquad (f \circledast g)(x) = \int_a^x f\big[K(x,t)\big]\, g(t)\, d\phi(t),$$

and if

$$(2.1.5) \qquad 1 \circledast f = f \circledast 1 \in C(a,\infty),$$

for every $f \in C(a,\infty)$, then

(a) ϕ is continuous and strictly monotonic in $a \leq x < \infty$,

(b) the function $K(x,t)$ can be expressed in the form

$$(2.1.6) \qquad K(x,t) = \phi^{-1}\big(\phi(x)-\phi(t)\big),$$

where ϕ^{-1} is the inverse of ϕ, and

(c) $C(a,\infty)$ is a convolution ring with respect to the convolution as ring multiplication.

In this theorem $C(a,\infty)$ denotes the set of all continuous real functions of a real variable $x \geq a > -\infty$. We use the notation $*_\phi$ for the convolution

$$(2.1.7) \qquad (f *_\phi g)(x) = \int_a^x f\left[\phi^{-1}\big(\phi(x)-\phi(t)\big)\right] g(t)\, d\phi(t),$$

in order to avoid confusion in later discussions; for $\phi(x) = x$ we denote the convolution simply by $*$. If we let $\odot$ denote composition of functions, i.e. $(f \odot g)(x) = f\big(g(x)\big)$, then it can be noted that for ϕ monotone and continuous the mapping $f \to f \odot \phi^{-1}$ plays an important role inasmuch as it carries the (Laplace) convolution ring of functions $C(0,\infty)$ with the multiplication $*$ into the generalized convolution ring of function $C(a,\infty)$ with the multiplication $*_\phi$;

that is,

$$(2.1.8) \qquad (f \odot \phi^{-1}) * (g \odot \phi^{-1}) = (f *_\phi g) \odot \phi^{-1}.$$

These results show that the better behaved equations, in one sense, of the form

$$(2.1.9) \qquad \int_a^x k(x,t) \, f(t) \, d\phi(t) = g(x),$$

are those which are equivalent to the equation $k * f = g$. In regard to the equivalent forms we list a few cases of particular interest, some of which appeared in Chapter 1.

Example 1. $\phi(t) = t^a$, $t \overset{\ge}{=} 0$, $a > 0$,

$$(2.1.10) \qquad (k *_\phi f)(x) = \int_0^x k\left((x^a - t^a)^{1/a}\right) f(t) \, at^{a-1} \, dt, \qquad x \overset{\ge}{=} 0,$$

or in the special case $a = 2$ which is often encountered

$$(2.1.11) \qquad (k *_\phi f)(x) = \int_0^x k\left((x^2 - t^2)^{1/2}\right) f(t) \, 2t \, dt, \qquad x \overset{\ge}{=} 0.$$

Example 2. $\phi(t) = \log t$, $t \overset{\ge}{=} 1$,

$$(2.1.12) \qquad (k *_\phi f)(x) = \int_1^x t^{-1} k(x/t) f(t) \, dt, \qquad x \overset{\ge}{=} 1.$$

Example 3. $\phi(t) = -\log t$, $0 < t \overset{\le}{=} 1$,

$$(2.1.13) \qquad (k *_\phi f)(x) = \int_x^1 t^{-1} k(x/t) f(t) \, dt, \qquad 0 < x \overset{\le}{=} 1.$$

It should be noted that in Examples 2 and 3 if $\phi(t)$ is replaced by $\phi(t/a)$, then the intervals are oriented about $\underline{a}$ instead of 1; thus we should replace 1 by $\underline{a}$ as the limit of integration and $k(x/t)$ by $k(ax/t)$ in the integrand. In Example 1 if $a < 0$, the lower limit then reads ∞; a particular case of this follows.

Example 4. $\phi(t) = t^{-1}$, $0 < t < \infty$,

$$(2.1.14) \qquad (k *_\phi f)(x) = \int_x^\infty k\left(xt(t-x)^{-1}\right) t^{-2} f(t) \, dt, \qquad x > 0.$$

We give below some other examples which may be of interest.

Example 5. $\phi(t) = \log(1+t)$, $t \gtreqless 0$,

$$(2.1.15) \qquad (k *_\phi f)(x) = \int_0^x k\left((x-t)(1+t)^{-1}\right) f(t) \, (1+t)^{-1} \, dt, \qquad x > 0,$$

which, after changes of variables, could also be written as

$$(2.1.16) \qquad (k *_\phi f)(x-1) = \int_1^x u^{-1} k(x/u-1) \, f(u-1) \, du, \qquad x > 1.$$

Example 6. $\phi(t) = -\log(1-t)$, $0 \lesseqgtr t < 1$,

$$(2.1.17) \qquad (k *_\phi f)(x) = \int_0^x k\left((x-t)(1-t)^{-1}\right) f(t) \, (1-t)^{-1} \, dt, \qquad 0 \lesseqgtr x < 1,$$

and as in Example 5,

$$(2.1.18) \qquad (k *_\phi f)(1-x) = \int_x^1 u^{-1} k(1-x/u) \, f(1-u) \, du, \qquad 0 < x \lesseqgtr 1.$$

Example 7. $\phi(t) = \sinh^{-1} t$, $t \gtreqless 0$,

(2.1.19) $(k *_\phi f)(x) = \int_0^x (1+t^2)^{-1/2} k\left(x(1+t^2)^{1/2}-t(1+x^2)^{1/2}\right) f(t)\, dt,$

$$x \gtreqless 0.$$

Example 8. $\phi(t) = \log(t+a) - \log(t-a),\quad t > a > 0,$

(2.1.20) $(k *_\phi f)(x) = \int_x^\infty 2a(t^2-a^2)^{-1} k\left((xt-a^2)(t-x)^{-1}\right) f(t)\, dt,\quad x > a > 0$

Example 9. $\phi(t) = \log(a+t) - \log(a-t),\quad 0 < t < a,$

(2.1.21) $(k *_\phi f)(x) = \int_0^x 2a(a^2-t^2)^{-1} k\left(a^2(x-t)(a^2-xt)^{-1}\right) f(t)\, dt,$

$$0 < x < a.$$

Example 10. $\phi(t) = \arcsin t,\quad 0 \leqq t \leqq 1,$

(2.1.22) $(k *_\phi f)(x) = \int_0^x (1-t^2)^{-1/2} k\left(x(1-t^2)^{1/2}-t(1-x^2)^{1/2}\right) f(t)\, dt,$

$$0 \leqq x \leqq 1.$$

This last example is of a different type since the range of $\phi(t)$ is finite.

2.2 The theorem of Titchmarsh and uniqueness of solutions

We next turn our attention to the theorem of Titchmarsh [140:10.11], the importance of which is also discussed in the books of Berg [1] and Erdélyi [37] in connection with the formation of a field of quotients.

Theorem. If $k * f = 0$, then $k = 0$ or $f = 0$.

This result is, of course, equivalent to a theorem on the uniqueness

of solutions for the (Laplace) convolution equation $k * f = g$ and

consequently under the appropriate mappings the generalized convolutions

of our examples lead to uniqueness of solutions of the corresponding

equations.

Examples of the form

$$(2.2.1) \qquad \int_0^x k(t/x)\, f(t)\, dt = g(x), \qquad x \geq 0,$$

give us examples which do not come under this class of equations which

have unique solutions; this was illustrated by the developments of

Chambers and Sack [24, 25, 100] for Legendre kernels. The substitution

$u = t/x$ puts equation (2.2.1) into the format

$$(2.2.2) \qquad \int_0^1 k(u)\, f(ux)\, x\, du = g(x),$$

which shows that if, for example, $f(ux) = (ux)^m$ and $k(u) = p_n(u)$,

where the p_n belong to a set of polynomials which are orthogonal on

the interval $(0, 1)$, then at least for certain kernels we have

non-trivial solutions of the homogeneous equation (2.2.2). We note

also that an exponential substitution converts an equation of type

(2.2.1) to the form

$$(2.2.3) \qquad (k \overset{\infty}{*} f)(x) = \int_x^\infty k(t-x)\, f(t)\, dt = g(x), \qquad x \geq 0,$$

so that this convolution over an unbounded interval possesses quite

different properties than those for the (Laplace) convolution. The

algebra involving this $\overset{\infty}{*}$-convolution as multiplication is non-associative and non-commutative; the identity

$$(2.2.4) \qquad h \overset{\infty}{*} (k \overset{\infty}{*} f) = k \overset{\infty}{*} (h \overset{\infty}{*} f),$$

does, however, hold. One method of investigating uniqueness is to try to reduce the operator into appropriate factors where uniqueness is known; see, for example, Love [66, 67].

2.3 Associated integral transformations

Since the convolution $k * f$ is associated with the Laplace transformation, which we choose here in the form

$$(2.3.1) \qquad \hat{f} = L\{f\} = \int_0^\infty e^{-st} f(t) \, dt,$$

therefore, for those functions ϕ which have range $(0, \infty)$, it is not difficult to write down the integral transformation T_ϕ which is associated with the generalized convolution. In consideration of (2.1.8) we choose

$$(2.3.2) \qquad T_\phi\{f\} = L\{f \odot \phi^{-1}\},$$

in order to preserve the basic property

$$(2.3.3) \qquad T_\phi\{k \underset{\phi}{*} f\} = T_\phi\{k\} \, T_\phi\{f\}.$$

Referring to our numbered examples, we can list the following results.

Example 1. $\quad T_\phi\{f\} = \int_0^\infty \exp(-su^a) \, f(u) \, au^{a-1} \, du.$

Example 2. $\quad T_\phi\{f\} = \int_1^\infty u^{-s-1} \, f(u) \, du.$

Example 3. $T_\phi\{f\} = \displaystyle\int_0^1 u^{s-1} f(u)\, du.$

Example 4. $T_\phi\{f\} = \displaystyle\int_0^\infty e^{-s/u} f(u)\, u^{-2}\, du.$

Example 5. $T_\phi\{f\} = \displaystyle\int_0^\infty (1+u)^{-s-1} f(u)\, du.$

Example 6. $T_\phi\{f\} = \displaystyle\int_0^1 (1-u)^{s-1} f(u)\, du.$

Example 7. $T_\phi\{f\} = \displaystyle\int_0^\infty (u^2+1)^{-1/2}\bigl(u+(u^2+1)^{1/2}\bigr)^{-s} f(u)\, du.$

Example 8. $T_\phi\{f\} = 2a\displaystyle\int_a^\infty (u-a)^{s-1}(u+a)^{-s-1} f(u)\, du.$

Example 9. $T_\phi\{f\} = 2a\displaystyle\int_0^a (a-u)^{s-1}(a+u)^{-s-1} f(u)\, du.$

On inspection of these examples it is evident that the Mellin transformation is closely related to Examples 2, 3, 5, and 6, whereas the Laplace transformation is related to Examples 1 and 4; hence these transformations are well suited for application to the related integral equations.

Example 10 illustrates a case which cannot in general be associated with the Laplace transformation by means of (2.3.2), since the range of $\phi(t)$ is not $(0, \infty)$, but rather $(0, \pi/2)$. It can, however, be associated with the finite Laplace transformation

(2.3.4) $\qquad L_T\{f\} = \displaystyle\int_0^T e^{-st} f(t)\, dt = \int_0^\infty e^{-st} f(t)\, U(T-t)\, dt,$

where $T = \pi/2$. Thus (2.3.2) must be replaced by

(2.3.5) $\qquad T_\phi\{f\} = L_{\pi/2} \{f \circledast \phi^{-1}\},$

and then we have

(2.3.6) $\qquad T_\phi\{f\} = \int_0^1 e^{-s \arcsin u} f(u)(1-u^2)^{-1/2} du.$

2.4 Associated fractional integrals

Fractional integrals of the type

(2.4.1) $\qquad I^\alpha f(x) = \dfrac{1}{\Gamma(\alpha)} \int_0^x (x-t)^{\alpha-1} f(t)\, dt$

were discussed by Kober [63] including the uniqueness property that $I^\alpha f_1 = I^\alpha f_2$ implies $f_1 = f_2$; he also considered the analog with range of integration (x, ∞). The form is that of a (Laplace) convolution and thus the fractional integrals can be defined for the general convolutions. If we set

(2.4.2) $\qquad k\left[\phi^{-1}\big(\phi(x)-\phi(t)\big)\right] = \big(\phi(x)-\phi(t)\big)^{\alpha-1}/\Gamma(\alpha),$

we obtain

(2.4.3) $\qquad I_\phi^\alpha\, f(x) = \dfrac{1}{\Gamma(\alpha)} \int_0^x \big(\phi(x)-\phi(t)\big)^{\alpha-1} f(t)\, d\phi(t).$

The fractional integrals of the form

(2.4.4) $\qquad I_\phi^{\eta,\alpha}\, f(x) = \big(\phi(x)\big)^{-\eta-\alpha}\, I_\phi^\alpha\left[\big(\phi(x)\big)^\eta f(x)\right],$

which are usually referred to as the Erdélyi-Kober fractional integrals, can consequently be introduced. We illustrate the explicit forms for a few of our examples.

Example 1. $\quad I_\phi^\alpha \ f(x) = \dfrac{1}{\Gamma(\alpha)} \displaystyle\int_0^x (x^a - t^a)^{\alpha-1} \ f(t) \ at^{a-1} \ dt;$

(2.4.5) $\quad I_\phi^{\eta,\alpha} \ f(x) = \dfrac{x^{-a(\eta+\alpha)}}{\Gamma(\alpha)} \displaystyle\int_0^x (x^a - t^a)^{\alpha-1} \ f(t) \ at^{a\eta+a-1} \ dt$

$$= \frac{a}{\Gamma(\alpha)} \int_0^x t^{-1} \left((x/t)^a - 1\right)^{\alpha-1} (x/t)^{-a(\eta+\alpha)} \ f(t) \ dt.$$

Example 2. $\quad I_\phi^\alpha \ f(x) = \dfrac{1}{\Gamma(\alpha)} \displaystyle\int_1^x t^{-1} \ \log^{\alpha-1}(x/t) \ f(t) \ dt;$

2.4.6) $\quad I_\phi^{\eta,\alpha} \ f(x) = \dfrac{\log^{-\eta-\alpha}x}{\Gamma(\alpha)} \displaystyle\int_1^x t^{-1} \ \log^{\alpha-1}(x/t) \ \log^\eta t \ f(t) \ dt.$

Example 3. $\quad I_\phi^\alpha \ f(x) = \dfrac{1}{\Gamma(\alpha)} \displaystyle\int_x^1 t^{-1} \ \log^{\alpha-1}(t/x) \ f(t) \ dt;$

2.4.7) $\quad I_\phi^{\eta,\alpha} \ f(x) = \dfrac{\log^{-\eta-\alpha}(1/x)}{\Gamma(\alpha)} \displaystyle\int_x^1 t^{-1} \ \log^{\alpha-1}(t/x) \ \log^\eta(1/t) \ f(t) \ dt.$

Example 4. $\quad I_\phi^\alpha \ f(x) = \dfrac{x^{1-\alpha}}{\Gamma(\alpha)} \displaystyle\int_x^\infty (t-x)^{\alpha-1} \ t^{-\alpha-1} \ f(t) \ dt;$

2.4.8) $\quad I_\phi^{\eta,\alpha} \ f(x) = \dfrac{x^{1+\eta}}{\Gamma(\alpha)} \displaystyle\int_x^\infty (t-x)^{\alpha-1} \ t^{-1-\eta-\alpha} \ f(t) \ dt.$

It should be noted that this last fractional integral operator is usually denoted by $K_\phi^{\eta+1,\alpha}$ in the literature.

A number of generalizations of the fractional integral operators have been obtained by introducing an additional function into the integrand. An example of such a generalization with Bessel functions is given by Lowndes [68] who also obtains the necessary inversion formulas. Because of the basic identity for the exponential function the choice

$$(2.4.9) \qquad (k \odot \phi^{-1})(z) = e^{bz} z^{\alpha-1}/\Gamma(\alpha),$$

leads to the related fractional integral operator

$$(2.4.10) \qquad e^{b\phi(x)} I_\phi^\alpha \left(e^{-b\phi(x)} f(x) \right)$$

which has appeared, for example, in the works of Buschman [19], Habibullah [51], and Prabhakar [82, 85], for $\phi(x) = x$ in the form

$$(2.4.11) \qquad e^{bx} I_\phi^\alpha \left(e^{-bx} f(x) \right) = \frac{1}{\Gamma(\alpha)} \int_0^x e^{b(x-t)} (x-t)^{\alpha-1} f(t) \, dt.$$

We note that in connection with Example 2 the analog is

$$(2.4.12) \qquad x^b I_\phi^\alpha \left(x^{-b} f(x) \right) = \frac{1}{\Gamma(\alpha)} \int_1^x t^{-1} (x/t)^b \log^{\alpha-1}(x/t) f(t) \, dt.$$

Integral operators of the form

$$(2.4.13) \qquad \frac{1}{\Gamma(\alpha)} \int_a^x (x-t)^{\alpha-1} f(t) \, dt$$

for $a \stackrel{>}{=} 0$ and of the form

$$(2.4.14) \qquad \frac{1}{\Gamma(\alpha)} \int_x^b (t-x)^{\beta-1} f(t) \, dt$$

for $b \stackrel{<}{=} \infty$ have been introduced by several authors. These are the appropriate integral operators for treating equations of the form

$$(2.4.15) \qquad \int_a^x k(x-t)\, f(t)\, dt = g(x), \qquad a \overset{\geq}{=} 0,$$

and

$$(2.4.16) \qquad \int_x^b k(t-x)\, f(t)\, dt = g(x), \qquad b \overset{\leq}{=} \infty,$$

respectively. In the Inversion Tables we shall denote the operators (2.4.13) and (2.4.14), respectively, by $I_{x,a}^{\alpha}$ and $K_{x,b}^{\alpha}$ or for the generalized convolution by $I_{\phi,a}^{\alpha}$ and $K_{\phi,b}^{\alpha}$. Further, operators analogous to (2.4.4) can be introduced and they can be denoted by $I_{\phi,a}^{n,\alpha}$ and $K_{\phi,b}^{n,\alpha}$. As has been pointed out by Love [66, 67] and by others the equations with $a > 0$ or with $b < \infty$ can also be treated by truncations of the respective solutions of the corresponding equations with $a = 0$ provided $g(x) = 0$ for $0 < x < a$ or with $b = \infty$ provided $g(x) = 0$ for $b < x$.

2.5 Simple kernel variations

It is apparent that the equation of the form

$$(2.5.1) \qquad \int_0^x e^{a(x-t)}\, k(x-t)\, f(t)\, dt = g(x)$$

can be replaced by the equation

$$(2.5.2) \qquad \int_0^x k(x-t)\, F(t)\, dt = G(x)$$

where $F(t) = e^{-at} f(t)$ and $G(x) = e^{-ax} g(x)$. Because of this the

Inversion Tables usually do not contain both forms.

Simple cases which involve an adjustment of the constants can also be treated; for example, for $k*f = g$, if $k = ck_1$ we can replace f by cf and solve for cf in terms of g. If $k(t) = k_1(at)$ a change of variables can be used to reduce the problem to

$$(2.5.3) \qquad \int_0^x k_1(x-t)\ F(t)\ dt = G(x)$$

where $F(t) = f(t/a)$ and $G(x) = ag(x/a)$. These adjustments allow us to extend the use of the tables.

If the integral operator can be factored, for example into fractiona integrals, these factors can sometimes be rearranged to give different useful forms. The example in Section 4.3 illustrates this from the viewpoint of Mikusiński operators.

In Section 3.7 an example is given which illustrates how the introduction of generalized functions into the form of the solution can be used to overcome the restrictions of the form

$$(2.5.4) \qquad D_x^r\ g(x)\Big|_{x=0} = 0$$

which appear in the Inversion Tables.

CHAPTER 3

METHODS AND EXAMPLES

3.1 Rodrigues' formula

Except for the independent interest because of the very elementary ideas involved, this method should be merely included under section 4.5, Fractional integrals. It was suggested by Erdélyi [38] that whenever a Rodrigues' formula is available for the kernel function, then merely repeated differentiations and integrations may well suffice to solve the equation. As an example he considered the Legendre polynomial kernel with the equation in the form

$$(3.1.1) \qquad \int_1^x P_n(x/t)\, f(t)\, dt = g(x), \qquad g(1) = 0, \quad x > 1.$$

Since

$$(3.1.2) \qquad P_n(z) = (2^n n!)^{-1} (d/dz)^n \{(z^2-1)^n\},$$

the equation can be rewritten in the form

$$(3.1.3) \qquad \left(\frac{d}{dx}\right)^n \int_1^x (x^2-t^2)^n\, t^{-n}\, f(t)\, dt = 2^n n!\, g(x);$$

the equivalence to the given equation follows from the Leibniz rule. Next, an n-fold integration produces the equation

$$(3.1.4) \qquad \int_1^x (x^2-t^2)^n\, t^{-n}\, f(t)\, dt = 2^n n \int_1^x (x-t)^{n-1}\, g(t)\, dt,$$

which has a simpler kernel. Successive differentiations with respect to x^2 will simplify this further to

$$(3.1.5) \qquad n! \int_1^x t^{-n} f(t) \, dt = 2^n n \left(\frac{d}{d(x^2)}\right)^n \int_1^x (x-t)^{n-1} g(t) \, dt,$$

which can be solved by one additional differentiation with respect to x. Hence we can rewrite the solution in the format

$$(3.1.6) \qquad f(x) = \frac{x^{n+1}}{(n-1)!} \left(\frac{1}{x}\frac{d}{dx}\right)^{n+1} \int_1^x (x-t)^{n-1} g(t) \, dt$$

$$= x^{n+1} \omega^{n+1} I^n g(x),$$

where $\omega = x^{-1} d/dx$ and I^n denotes the n-fold iteration of the integral on the interval $(1, x)$; thus conditions on g for the validity of the solution can now readily be seen.

Erdélyi also points out that fractional integrals are needed for Chebyshev polynomials, Gegenbauer polynomials, and associated Legendre function kernels; C. Singh [110] uses this idea for Jacobi polynomial kernels.

There are other cases to which the elementary method can be extended. For example, if the kernel can be written in the form

$$(3.1.7) \qquad k(z) = c(n,k,\alpha) \, \Omega^n\left(z^\beta(z^\alpha-1)^{n+k}\right)$$

where $\Omega = z^{\gamma+1} d/dz$ and $c(n,k,\alpha)$ is independent of z, then the solution can be expressed in operators similar to Ω and to its inverse operator which involves integration over the interval $(1, x)$. A case for α equal to an integer appears in R.P. Singh [112].

3.2 Resolvent kernel

The basic idea is to consider an equation of the form

$$(3.2.1) \qquad \int_a^x k(x,t)\ f(t)\ dt = g(x),$$

and a proposed solution of the form

$$(3.2.2) \qquad \int_a^t k_1(t,u)\ \Lambda\ g(u)\ du = f(t),$$

in which Λ denotes some suitable linear operator (for example, a differentiation operator). We simply substitute the proposed solution into the equation and interchange the order of integration. Then we try to show that

$$(3.2.3) \qquad \int_a^x \Lambda\ g(u) \int_u^x k(x,t)\ k_1(t,u)\ dt\ du$$

reduces to $g(x)$. If the inner integral is a suitable function, then the reduction can often be verified by suitable differentiations and integrations.

This is the method applied by Ta Li [64] and much of his work involved the evaluation of the integral of the product of the kernels. In certain cases this integral may be known from tables, in which case the method is simpler to apply. Various methods, including the use of integral transforms, have been used to evaluate these key integrals, which are, of course, of convolution type. This method has been more useful than it might appear; if an inversion integral is conjectured which involves a similar kernel and is such that the integral involving the kernels is known, and if further there are parameters which can be adjusted to provide simplifications, then a simple expression can often be obtained by adjusting these parameters and by appropriately choosing the operator Λ.

As an example we recall the problem of Ta Li [64]

$$(3.2.4) \qquad \int_{x}^{1} (t^2-x^2)^{-1/2} T_n(t/x) \ f(t) \ dt = g(x), \quad g(1) = 0, \ 0 < x < 1,$$

where the proposed solution is

$$(3.2.5) \qquad f(t) = -\frac{2}{\pi} \int_{t}^{1} v^{1-n}(v^2-t^2)^{-1/2} T_{n-1}(t/v) \ d[v^n g(v)].$$

This leads to the problem of evaluating the integral

$$(3.2.6) \qquad \int_{x}^{v} \left((t^2-x^2)(v^2-t^2)\right)^{-1/2} T_n(t/x) \ T_{n-1}(t/v) \ dt,$$

by some method, which consequently reduces the problem to showing merely that

$$(3.2.7) \qquad -x^{-n} \int_{x}^{1} d[v^n g(v)]$$

reduces to $g(x)$.

3.3 Laplace transformation

This is a relatively old method which was discussed for integro-differential equations by Doetsch [31], who included Abel's equation as a specific illustration of how to treat the convolution integral equation of the first kind. The method is based upon the properties of the convolution integral

$$(3.3.1) \qquad (k * f)(x) = \int_0^x k(x-t) \, f(t) \, dt = g(x) \qquad\qquad x \geq 0,$$

and the fact that the Laplace transformation of this convolution is given by

$$(3.3.2) \qquad L\{k * f\} = L\{k\} \, L\{f\}.$$

We choose the Laplace transformation in the form

$$(3.3.3) \qquad \hat{f} = L\{f\} = \int_0^\infty e^{-st} \, f(t) \, dt,$$

in order to correspond to the tables [27], [42], and [89]. Thus the integral equation $k * f = g$ is transformed into the algebraic equation $\hat{k}\hat{f} = \hat{g}$ so that $\hat{f} = \hat{g}/\hat{k}$. However, $1/\hat{k}$ is never a Laplace transform when k is one; thus one can never merely write $\hat{f} = \hat{k}_1\hat{g}$ and invert. One simple manipulation [81] which is useful to try is to choose a polynomial $p_n(s)$, or some other suitable function, and rewrite the equation in the form $\hat{f} = \hat{k}_2(p_n g)$ where $\hat{k}_2 = 1/(p_n \hat{k})$ is actually a Laplace transform and $g^{(i)}(0) = 0$ for $i = 0,1,2,\cdots,n-1$. In this case the result of the inversion gives us the particularly simple form of a convolution

$$(3.3.4) \qquad f(x) = \int_0^x k_2(x-t) \left(p_n(D)g(t) \right) \, dt,$$

where $D = d/dt$. One of a number of alternative forms,

$$(3.3.5) \qquad f(x) = p_n(D) \int_0^x k_2(x-t) \, g(t) \, dt,$$

results from merely the regrouping of the transforms to the form $p_n(\hat{k}_2\hat{g})$ before inverting. Often $p_n(s) = s^n$ is suitable, but other polynomials should not be overlooked; Widder [143] uses $p_n(s) = (s-1)^2$ with the Laguerre kernel. As a matter of fact, H.M. Srivastava [124] observed that if we have

$$(3.3.6) \qquad \hat{k}(s) = \left\{ (s-\alpha)^n \, \hat{k}_3(s) \right\}^{-1}$$

where $\hat{k}_3(s)$ is the Laplace transform of $k_3(t)$, continuous for $t \overset{\geq}{=} 0$, then the convolution integral equation (3.3.1) has for its solution,

$$(3.3.7) \qquad f(x) = \int_0^x k_3(x-t) \left\{ (D_t - \alpha)^n \, g(t) \right\} \, dt, \qquad D_t = d/dt,$$

provided that $g(x) \; \varepsilon \; C^n$ for $0 \overset{\leq}{=} x < \infty$ and $g^{(i)}(0) = 0$ $i = 0, 1, \cdots, n-1$.

The formalism along with the available tables [27], [30], [42], [73] and [74], [80], or [89] makes this a convenient method. First, a very simple example shows how the method is applied to a specific case. Consider

$$(3.3.8) \qquad \int_0^x \sin(x-t) \, f(t) \, dt = g(x)$$

where $\hat{g}$ is assumed to exist. Since $L\{\sin t\} = (s^2+1)^{-1}$, we have

$$(3.3.9) \qquad \hat{f}(s) = (s^2+1) \, \hat{g}(s).$$

However, assuming that the Laplace transform $L\{D^2 g\}$ exists, we have $s^2\hat{g} = L\{D^2 g\}$ if and only if $g(0) = g'(0) = 0$, and under these conditions we finally obtain

(3.3.10) $f(t) = g''(t) + g(t)$.

It should be noted that the formalism has given us the correct form
for the solution even in cases where $\hat{g}$ fails to exist. For example,
if $g(t) = \exp(t^2)$, the result can be verified by direct substitution.

As another example, from the tables we have the pair of relations

(3.3.11) $L\{J_0(2(at)^{1/2})\} = s^{-1} e^{-a/s}$, $L\{I_0(2(at)^{1/2})\} = s^{-1} e^{a/s}$;

several authors have noted that these are useful for the two Bessel
function kernels. This example illustrates how the process also gives
us a solution to the equation with the I_0-kernel as well as with the
J_0-kernel; thus we often can solve a pair of such equations at the
same time when applying integral transforms. Because of the symmetry
in this case, I_0 and J_0 can merely be interchanged. The solution
form appears later in this section, see (3.3.12) and (3.3.15).

The relation to the resolvent kernel method can be seen by writing
$k * f = g$ and $f = k_1 * \Lambda g$ so that substitutions give us
$\Lambda g * (k * k_1) = \Lambda g * k_2$, which must be shown to reduce to g. The
evaluation of $k * k_1$ has sometimes actually been completed by taking
the transform of this convolution product, simplifying, and then
inverting.

A somewhat different point of view in the manner of the use of
tables was taken by Colombo [29], Parodi [77], and Poli [81]. As an
illustration we again consider the Bessel kernel

(3.3.12) $\int_0^x J_0[(x^2-t^2)^{1/2}] f(t)\, dt = g(x)$.

From the general substitution formula [42:4.1(37)] we obtain

(3.3.13) $\qquad (p^2+1)^{-1/2} \, \hat{f}\!\left((p^2+1)^{1/2}\right) = \hat{g}(p),$

and hence

(3.3.14) $\qquad \hat{f}(s) = s \, \hat{g}\!\left((s^2-1)^{1/2}\right),$

so that from [40:5.1(13)] we obtain the result

(3.3.15) $\qquad f(x) = \displaystyle\int_0^x I_0\!\left[(x^2-t^2)^{1/2}\right] D^2 g(t) \, dt.$

where it is assumed that $g(0) = g'(0) = 0$.

3.4 Mellin and other transformations

For the Mellin transformation the method is essentially the same
as for the Laplace transformation and the appropriate forms of the
convolution are those of Examples 2 and 3. A two-way table of Mellin
transforms, using the definition

(3.4.1) $\qquad M\{f\} = \displaystyle\int_0^\infty x^{s-1} \, f(x) \, dx,$

is available [42:6,7], but the useful ones are only those for which
the functions are zero over part of their domain of definition; this
is the analog which results from our choice of the Laplace transformation
being one-sided instead of two-sided.

A simple example should suffice for illustration; we consider

(3.4.2) $\qquad \displaystyle\int_x^1 (t^2-x^2)^{-1/2} \, f(t) \, dt = g(x), \quad g(1) = 0, \quad 0 < x \leqq 1.$

If we rewrite this equation in the form of an ordinary Mellin convolution [42:6.1(14)] we have

(3.4.3) $\quad \int_0^\infty t^{-1}\left\{(1-x^2/t^2)^{-1/2}\ U(1-x/t)\right\}\left\{f(t)\ U(1-t)\right\}\ dt = g(x)\ U(1-x),$

where U denotes the Heaviside unit function, defined by

(3.4.4) $\qquad U(x) = \begin{cases} 0, \text{ if } x < 0 \\ 1, \text{ if } x > 0. \end{cases}$

Hence we have

(3.4.5) $\quad \left\{(1-x^2)^{-1/2}\ U(1-x)\right\} *_M \left\{f(x)\ U(1-x)\right\} = \left\{g(x)\ U(1-x)\right\},$

where $*_M$ denotes the Mellin convolution. Equation (3.4.5) can now be transformed by using [42:6.2(31)] into

(3.4.6) $\qquad \dfrac{\Gamma(1/2)\Gamma(s/2)}{2\Gamma(s/2+1/2)}\ M\left\{f(x)\ U(1-x)\right\} = M\left\{g(x)\ U(1-x)\right\}.$

Solving for the transform of f we have

(3.4.7)
$$M\left\{f(x)\ U(1-x)\right\} = \dfrac{\Gamma(s/2-1/2)\Gamma(1/2)}{\pi\Gamma(s/2)}\ (s-1)\ M\left\{g(x)\ U(1-x)\right\}$$
$$= (-2/\pi)\ M\left\{x^{-1}(1-x^2)^{-1/2}U(1-x)\right\}\left(-(s-1)\right)M\left\{g(x)U(1-x)\right\}.$$

Thus from [42:6.1(14)] using also [42:6.1(31),(9), and (4)] we finally obtain

(3.4.8) $\qquad f(x) = -\dfrac{2}{\pi x}\int_x^1 (t^2-x^2)^{-1/2}\ t\ g'(t)\ dt, \qquad 0 < x < 1,$

where conditions must now be placed on g' so that the integral exists.

It should be noted that the integral transformations which were introduced in Examples 2 and 3 of Chapter 2 are more directly applicable to such problems involving kernels of the form $k(x/t)$ with integration extending over the intervals $(1,x)$ and $(x,1)$, since the Heaviside function is incorporated within the definition of the transformation. In fact, in the example just given it can be seen in (3.4.5) and (3.4.6) that we actually have used merely the transformation of Example 3.

The same general idea would apply to any integral transformation where a convolution is known; however, it may not be very useful unless there are extensive tables available. Also, unless there is a Titchmarsh theorem for the convolution, the difficult problem of uniqueness would remain. The Fourier convolution, for example, has non-trivial divisors of zero; a simple example of this can be obtained from the known formulas $[42:3.2(3),(4)]$.

3.5 Fractional integrals

A basic idea involving the application of fractional integrals is to apply some such operator I to $k *_\phi f = g$ so that we have $k_1 *_\phi f = g$; this method is productive provided the kernel k_1 is simpler than k. From the results of section 2.4 it is readily seen that this can often be interpreted merely as convolution with the function associated with I, that is, as $h *_\phi k *_\phi f = h *_\phi g$ so that $k_1 = h *_\phi k$. It is also clear then that such methods are closely allied with integral transformations. Inverting the fractional integral is, of course, itself a problem involving the solution of an intetral equation of the kind under discussion.

One of the most effective applications of this method is contained in the work of Love [66, 67]. He shows that the four equations (1.9.1) through (1.9.4), involving the Gauss hypergeometric function in the kernel, of which our Examples 5 and 6 are only special cases, can be rewritten with the operator factored into a product involving fractional integrals and power function multipliers. Uniqueness of solution is also obtained. For kernels involving Kummer's function, a similar factorization involves exponential function multipliers.

We first illustrate the method by the simple example

$$(3.5.1) \qquad \frac{1}{\Gamma(c)} \int_0^x (x-t)^{c-1} {}_1F_1(a;c;x-t) \, f(t) \, dt = g(x) \qquad x > 0.$$

In view of [40:13.1(94)] and the elementary identity

$$(3.5.2) \qquad {}_1F_1(a;a;x) = e^x,$$

we have

$$(3.5.3) \qquad I^{a-c} x^{c-1} {}_1F_1(a;c;x)/\Gamma(c) = x^{a-1} e^x/\Gamma(a).$$

Thus the equation (3.5.1) can be rewritten in the form

$$(3.5.4) \qquad \frac{1}{\Gamma(a)} \int_0^x e^{x-t}(x-t)^{a-1} \, f(t) \, dt = I^{a-c} g(x).$$

Next, after splitting up the exponential function, we have

$$(3.5.5) \qquad I^a\left(e^{-x}f(x)\right) = e^{-x} I^{a-c} g(x),$$

and hence

$$(3.5.6) \qquad f(x) = e^x I^{-a} e^{-x} I^{a-c} g(x),$$

from which the needed conditions on $g(x)$ for a solution to exist can now be obtained.

As another example we consider the equation

$$(3.5.7) \qquad \frac{1}{\Gamma(c)} \int_0^x t^{-1}(1-t/x)^{c-1}(t/x)^{\gamma-1} {}_2F_1(a,b;c;1-t/x) \ f(t) \ dt = g(x).$$

In terms of the Erdélyi-Kober operator (2.4.4), with $\phi(x) = x$, this equation can be rewritten as

$$(3.5.8) \qquad I^{\gamma+c-a-b,b} \ I^{\gamma,c-b} \ f = g$$

wherein use has been made of Euler's formula [41:2.1(10)].

Now the problem is to invert these fractional integrals; if we make use of the known properties

$$(3.5.9) \qquad I^{\eta,\alpha} \ I^{\eta+\alpha,\beta} = I^{\eta,\alpha+\beta}, \quad I^{\eta,0} = I^0,$$

and choose the operators so as to simplify the kernel, we shall obtain

$$(3.5.10) \qquad f(x) = I^{\gamma+c-b,b-c} \ I^{\gamma+c-a,-b} \ g(x),$$

where, of course, fractional integrals of negative order are interpreted in terms of differentiations of positive integral order and fractional integrals of positive order. Since these generalized operators can also be written in simple convolution form, that is, for $\phi(x) = x$

$$(3.5.11) \qquad I^{\eta,\alpha} = x^{-\eta-\alpha} \ I^\alpha \ x^\eta,$$

we have the alternate form

$$(3.5.12) \qquad f(x) = x^{-\gamma} \ I^{b-c} \ x^a \ I^{-b} \ x^{\gamma+c-a} \ g(x),$$

for the solution of the integral equation (3.5.7).

3.6 Mikusiński operators

This method is a direct extension of the Laplace transformation method; it also includes many of the aspects of fractional integrals. The advantage is that we extend the ring of functions to the set of operators (convolution quotients) which is a field; hence the operator equation $kf = g$, for $k \neq 0$, always possesses a solution g/k which is unique. The disadvantage is that we are led to the rather difficult problem of determining when a particular quotient of operators would correspond to a function from our original ring. In fact, except for very simple forms such as rational functions of the differentiation operator, the problem is open.

As an example of the technique we consider

$$(3.6.1) \qquad \int_0^x \cos(x-t)\, f(t)\, dt = a + bt.$$

The operator form of the equation is

$$(3.6.2) \qquad s(s^2+1)^{-1} f = as^{-1} + bs^{-2},$$

and the operator solution can be written as

$$(3.6.3) \qquad f = a + bs^{-1} + as^{-2} + bs^{-3}.$$

Since the operator s^{-c} corresponds to a function if and only if $\mathrm{Re}(c) > 0$, the solution is a function if and only if $a = 0$, when it is given by

$$(3.6.4) \qquad f(t) = b + bt^2/2.$$

If $a \neq 0$, the solution can be represented as a generalized function involving the "Dirac δ-function", or impulse, in the form

$$(3.6.5) \qquad f(t) = a\delta(t) + b + at + bt^2/2.$$

In general, those kernels with operator representations which are rational functions of the differentiation operator s are particularly suitable for this treatment, just as in the case of kernels whose Laplace transform is a rational function.

As in the integral transformation method, we often have different appearing solutions depending upon the factorization used for the operators before their representation as convolutions of functions. The table of Mikusiński and Ryll-Nardzewski [72] is useful here to obtain information about solutions in various function classes.

3.7 Other methods

The only other methods found in the literature which seem to show some promise are the partial differential equation method of Mackie [69], and the method of Suschowk [138], however, they do not seem to have been pursued further.

Mackie [69] notes that the equation

$$(3.7.1) \qquad \phi(r_0, r_0) = \frac{1}{2} \int_0^{2r_0} (r/r_0)^N \, P_{N-1}(r/2r_0) \, g(r) \, dr$$

arises from the application of Riemann's method to obtain the solution $\phi(r_0, t_0)$, for $0 \leq t_0 \leq r_0$, of the partial differential equation

$$(3.7.2) \qquad \frac{\partial^2 \phi}{\partial r^2} + \frac{2N}{r} \frac{\partial \phi}{\partial r} - \frac{\partial^2 \phi}{\partial t^2} = 0,$$

where $2N$ need not be integral, with the Cauchy data

(3.7.3) $\phi = 0, \quad \dfrac{\partial \phi}{\partial t} = g(r)$ at t = 0.

With this in mind a new boundary value problem can be set up with

$\phi = f(r)$ on r = t and $\phi = -f(r)$ on r = -t which is then solved

by use of the Riemann function and hence g(r) obtained by use

of (3.7.3).

Simplification of the details is obtained by working with

characteristic coordinates. The integral equations which are solved

by this method are those of the form

(3.7.4) $f(x_0) = \displaystyle\int_0^{x_0} R(x,x;x_0,0)\, g(x)\, dx$

where $R(x,y;x_0,y_0)$ is the Riemann function of the equation

(3.7.5) $\dfrac{\partial^2 \phi}{\partial x \partial y} + c(x,y)\, \phi = 0, \qquad c(y,x) = c(x,y).$

The Cauchy data

(3.7.6) $\phi = 0, \quad \dfrac{\partial \phi}{\partial x} - \dfrac{\partial \phi}{\partial y} = 2g(x),$ on y = x > 0

is used to set up (3.7.4). From the boundary value problem in which

$\phi = f(x)$ on y = 0(x > 0) and $\phi = -f(x)$ on x = 0(y > 0) a

solution to (3.7.4) can be computed as

(3.7.7) $g(x_0) = f'(x_0) + \displaystyle\int_0^{x_0} \left[\dfrac{\partial}{\partial x_0} - \dfrac{\partial}{\partial y_0} \right] R(x,0;x_0,y_0) \Big|_{y_0 = x_0} f'(x)\, dx.$

The special cases of two Legendre kernels (1.4.3), (1.4.4) and one

Bessel kernel similar to (1.4.7) are displayed.

Suschowk [138] obtains the solution of the equation

(3.7.8) $\qquad u^{(k)}(t,r) = \int_0^{t-r} \left((t-\tau)^2 - r^2\right)^{-1/2} T_k\left((t-\tau)/r\right) f(\tau) \, d\tau$

where $t \geqq r > 0$, $f(0) = 0$, and T_k is a Chebyshev polynomial, which arises in the solution of the partial differential equation

(3.7.9) $\qquad U_{tt} - U_{rr} - r^{-1} U_r - r^{-2} U_{ww} = 0$

with the conditions

(3.7.10) $\quad U_0 = U_t = U_r = U_w = 0$ for $t^2 - r^2 \geqq 0$, $t \geqq 0$.

The solution forms are

(3.7.11) $\qquad f(t) = - \lim_{r\downarrow 0} r\, u_r^{(0)}(t,r)$, $\qquad t \geqq 0$

for $k = 0$ and

(3.7.12) $\qquad f(t) = \dfrac{1}{2^{k-1}(k-1)!} \dfrac{d^k}{dt^k} \left| \lim_{r\downarrow 0} r^k u^{(k)}(t,r) \right|$, $\quad t \geqq 0$

for $k = 1, 2, \cdots$; the method used is direct computation starting from (3.7.8) to obtain (3.7.11) and (3.7.12).

The following two methods should be mentioned, although they do not seem to have been applied to any extent to the cases in which the kernels are special functions.

An algebra of formal power series has been introduced by Sandy Grabiner [49] in order to consider several kinds of convolution integral equations.

In some cases the equations of the first kind can be converted to equations of the second kind, for example see Section 6 of Chapter 2 of Hochstadt [55]. Assuming that g and k satisfy suitable conditions of differentiability and that $k(x,x)$ is defined and not zero, then

(0.1) can be converted to the equation of the second kind

(3.7.13)
$$f(x) + \int_a^x k_1(x,t)\, f(t)\, dt = g_1(x)$$

where

(3.7.14)
$$k_1(x,t) = \big(k(x,x)\big)^{-1} \frac{\partial}{\partial x} k(x,t), \qquad g_1(x) = \big(k(x,x)\big)^{-1} g(x).$$

Thus for the convolution equation $k*f = g$, $k(0)$ needs to be defined and not zero at the origin. Even if $k(0) = 0$ the conversion can be carried out provided repeated differentiation is possible and some derivative of k is defined and not zero at the origin.

CHAPTER 4

MISCELLANEOUS RESULTS

4.1 Some immediate results from tables of integral transforms

An inspection of tables of Laplace transformations leads to the following inversion pairs.

(I) For the polynomials $A_n(x)$ defined by

$$(4.1.1) \quad A_0(x) = 1, \quad A_{n+1}(x) = \int_0^x A_n(t) \, dt - (n+1)^{-1} A_n(x),$$

we can use [42:5.2(15)] to obtain the solution of the equation

$$(4.1.2) \quad \int_0^x A_n(x-t) \, f(t) \, dt = g(x), \quad g^{(m)}(0) = 0 \quad \text{for} \quad 0 \leq m \leq n,$$

in the form

$$(4.1.3) \quad f(x) = (-1)^n \, n! \int_0^x k(x-t) \, g^{(n+1)}(t) \, dt,$$

where

$$(4.1.4) \quad k(u) = e^u * e^{2u} * \cdots * e^{nu};$$

this last equation (4.1.4) can also be written in the form

$$(4.1.5) \quad k(u) = \sum_{m=1}^{n} c_m \, e^{mu}, \quad \text{with} \quad c_m = \prod_{j=1, j \neq m}^{n} (m-j)^{-1}.$$

(II) From [42:4.11(6),(7)] we easily obtain the solution of

$$(4.1.6) \quad \int_0^x P_n\big(\cosh(x-t)\big) \, f(t) \, dt = g(x), \quad g(0) = g'(0) = 0,$$

in the form

(4.1.7)
$$f(x) = \int_0^x P_{n-1}\left(\cosh(x-t)\right)(D_t^2 - n^2)\, g(t)\, dt.$$

(III) From [42:5.9(6), (8)] where the bracket notation, [x], is used to indicate the greatest integer not exceeding x, we find that the integral equation

(4.1.8)
$$\int_0^x (-1)^{[(x-t)/(2a)]}\, f(t)\, dt = g(x), \quad g(0) = g'(0) = 0,$$

has the solution

(4.1.9)
$$f(t) = \int_0^x \left(2[(x-t)/2a]+1\right)\, g''(t)\, dt.$$

(IV) From [42:4.26(6), 5.7(1)] we can consider the equation

(4.1.10)
$$\int_0^x \nu(t-x,a)\, f(t)\, dt = g(x), \quad g^{(m)}(0) = 0 \quad \text{for} \quad 0 \overset{\leq}{=} m < a + 2,$$

which is readily seen to have the solution given by

(4.1.11)
$$f(x) = \int_0^x \left(\psi(1) - \log(x-t)\right) I^{-a-2}\, g(t)\, dt.$$

(V) A rather special case of the generalized hypergeometric function obtained by taking m = 1, n = 0 in [42:5.21(3)] leads to the equation

(4.1.12)
$$\frac{1}{\Gamma(k\sigma)} \int_0^x (x-t)^{k\sigma-1}\, {}_1F_k\left[a; \sigma, \sigma+1/k, \cdots, \sigma+(k-1)/k; \left(\lambda(x-t)/k\right)^k\right]$$
$$\cdot f(t)\, dt = g(x),$$

which transforms into the simple relation

(4.1.13)
$$\left(s^k - \lambda^k\right)^{-\sigma} \hat{f} = \hat{g},$$

for $\mathrm{Re}(\sigma) > 0$, $\mathrm{Re}(s) > \mathrm{Re}(\lambda)$, so that

(4.1.14)
$$\hat{f} = (s^k - \lambda^k)^{\sigma} \hat{g}.$$

Consequently, $f(t)$ can be obtained in terms of a differential operator of order $k\sigma$ applied to g; thus $g^{(m)}(0) = 0$ for $0 \leq m < k\sigma$ will be needed. If we use the fact that the operator $(s - \lambda)^{-\nu}$ is equivalent to the operation $e^{\lambda x} I^{\nu} e^{-\lambda x}$ applied to a function, which has been discussed at the end of section 2.4, and if we let ω denote a primitive k-th root of unity we can write the solution in the form

(4.1.15)
$$f(x) = \prod_{m=1}^{k} \left(\exp(\lambda \omega^m x) \, I^{-\sigma} \, \exp(-\lambda \omega^m x)\right) g(x).$$

It should be noted that this indicates that the original operator could have been factored into fractional integrals and exponential multipliers.

(VI) From the known formulas [27,p.323] for periodic functions we obtain relations between solutions of certain equations. Let k denote a periodic function which satisfies $k(u+c) = -k(u)$, let $k_{1/2}$ denote the half-wave rectification of k, and let k_1 denote the full wave rectification of k. Further, let f, $f_{1/2}$, and f_1 represent solutions of the corresponding convolution equations. Since we have

(4.1.16)
$$\hat{k}_{1/2} = \left(1 - e^{-cs}\right)^{-1} \hat{k} \quad \text{and} \quad \hat{k}_1 = \mathrm{ctnh}(as/2) \, \hat{k},$$

we note that the transforms of the solutions must satisfy

(4.1.17)
$$\hat{f}_{1/2} = \left(1 - e^{-cs}\right) \hat{f} \quad \text{and} \quad \hat{f}_1 = \tanh(as/2) \, \hat{f}.$$

Consequently, we can write

(4.1.18) $\qquad f_{1/2}(t) = f(t) - f(t-c)\, U(t-c)$

and from the rewritten form

(4.1.19) $\qquad \hat{f}_1 = \hat{f} - 2s^{-1}\!\left(e^{cx}+1\right)^{-1} s\hat{f}$

we have, using [42:5.5(17)] and other properties,

(4.1.20) $\qquad f_1(t) = f(x) - \displaystyle\int_0^x \left(1-(-1)^{[(x-t)/c]}\right) f'(t)\, dt$

where [ ] again is the greatest integer notation, that is, the kernel

equals 0 for $2n < (x-t)/c < 2n + 1$ and equals 1 for

$2n + 1 < (x-t)/c < 2n + 2$.

4.2 Simplifications of generalized hypergeometric kernels

(VII) We note first that Clausen's theorem [41:4.3(1)], which
provides a factorization of a special case of $\,_3F_2$, can be used in
combination with [42:5.21(3)] to provide a convolution factorization
for a special case of a $\,_3F_{2+k}$ kernel. For $\text{Re}(\sigma) > \text{Re}(\rho) > 0$,
$k \stackrel{>}{=} 1$, this is

$$\frac{t^{k\sigma-1}}{\Gamma(k\sigma)}\,_3F_{2+k}\left(2a,2b,a+b;a+b+1/2,2a+2b,\sigma,\sigma+1/k,\cdots,\sigma+(k-1)/k;(\lambda x/k)^k\right)$$

(4.2.1)

$$= \left(x^{k\sigma-k\rho-1}/\Gamma(k\sigma-k\rho)\right)\,_2F_{1+k}\left(a,b;a+b+1/2,\sigma-\rho,\sigma-\rho+1/k,\cdots,\sigma-\rho+ \atop (k-1)/k;(\lambda x/k)^k\right)$$

$$* \left(x^{k\rho-1}/\Gamma(k\rho)\right)\,_2F_{1+k}\left(a,b;a+b+1/2,\rho,\rho+1/k,\cdots,\rho+(k-1)/k;(\lambda x/k)^k\right).$$

Consequently, convolution equations involving such $_3F_{2+k}$ kernels

for $k \overset{>}{=} 1$ can be reduced to solving successive equations which

involve $_2F_{1+k}$ kernels. In particular, for $k = 1$, the equation

$$(4.2.2) \quad \frac{1}{\Gamma(\sigma)} \int_0^x (x-t)^{\sigma-1} {}_3F_3\left(2a,2b,a+b;a+b+1/2,2a+2b,\sigma;\lambda(x-t)\right) f(t)dt = g(x)$$

can be reduced to successively solving the equations

$$(4.2.3) \quad \frac{1}{\Gamma(\rho)} \int_0^x (x-t)^{\rho-1} {}_2F_2\left(a,b;a+b+1/2,\sigma-\rho;\lambda(x-t)\right) h(t)\ dt = g(x),$$

and

$$(4.2.4) \quad \frac{1}{\Gamma(\sigma-\rho)} \int_0^x (x-t)^{\sigma-\rho-1} {}_2F_2\left(a,b;a+b+1/2,\rho;\lambda(x-t)\right)\ f(t)\ dt = h(x).$$

Formulas [41:4.3(8), (9)] lead to a similar form with a different

arrangement of parameters.

(VIII) From [41:4.3(2)] and [42:5.21(3)] we similarly obtain

$$\frac{t^{k\sigma-1}}{\Gamma(k\sigma)} {}_2F_{3+k}\left(a+b,a+b-1/2;2a,2b,2a+2b-1,\sigma,\sigma+1/k,\cdots,\sigma+(k-1)/k;(\lambda x/k)^k\right)$$

$$(4.2.5)$$

$$= \left(x^{k\sigma-k\rho-1}/\Gamma(k\sigma-k\rho)\right) {}_0F_{1+k}\left(;2a,\sigma-\rho,\sigma-\rho+1/k,\cdots,\sigma-\rho+(k-1)/k:(\lambda x/k)^k/4\right)$$

$$* \left(x^{k\rho-1}/\Gamma(k\rho)\right) {}_0F_{1+k}\left(;2b,\rho,\rho+1/k,\cdots,\rho+(k-1)/k;(\lambda x/k)^k/4\right),$$

for $\mathrm{Re}(\sigma) > \mathrm{Re}(\rho) > 1$ and $k \overset{>}{=} 1$. For $k = 1$ the equation

$$(4.2.6) \quad \frac{1}{\Gamma(\sigma)} \int_0^x (x-t)^{\sigma-1} {}_2F_4\left(a+b,a+b-1/2;2a,2b,2a+2b-1,\sigma;\lambda(x-t)\right) f(t)dt = g(x)$$

can be solved by treating successively

$$(4.2.7) \qquad \frac{1}{\Gamma(\sigma-\rho)} \int_0^x (x-t)^{\sigma-\rho-1} {}_0F_2\left(-;2a,\sigma-\rho;\lambda(x-t)/4\right) h(t) \, dt = g(x)$$

and

$$(4.2.8) \qquad \frac{1}{\Gamma(\rho)} \int_0^x (x-t)^{\rho-1} {}_0F_2\left(-;2b,\rho;\lambda(x-t)/4\right) f(t) \, dt = h(x).$$

Other formulas which provide factorizations might also be used; however, the resulting equations with ${}_2F_2$ and ${}_0F_2$ kernels which are special cases of the results of this section are not readily solvable. They do have the advantage in that fewer parameters are involved at each of the successive steps since ρ may be assigned a specific value for convenience.

(IX) The H-function of Fox [47] can be defined by the contour integral

$$H[x] = H_{p,q}^{m,n}\left[x \left| \begin{matrix} (a_1,A_1),\cdots,(a_p,A_p) \\ (b_1,B_1),\cdots,(b_q,B_q) \end{matrix} \right.\right]$$

$$(4.2.9) \qquad = \frac{1}{2\pi i} \int_L \frac{\displaystyle\prod_{j=1}^{m} \Gamma(b_j-B_j s) \prod_{j=1}^{n} \Gamma(1-a_j+A_j s)}{\displaystyle\prod_{j=m+1}^{q} \Gamma(1-b_j+B_j s) \prod_{j=n+1}^{p} \Gamma(a_j-A_j s)} x^s \, ds$$

where m, n, p, q are integers satisfying $0 \stackrel{<}{=} m \stackrel{<}{=} q$, $0 \stackrel{<}{=} n \stackrel{<}{=} p$, the A_j and B_j are all positive, the parameters are such that no poles of the integrand coincide, and the contour $\mathrm{Re}(s) = s_0$ separates the poles of one product from those of the other. If

$$(4.2.10) \qquad K = \sum_{j=1}^{n} A_j + \sum_{j=1}^{m} B_j - \sum_{j=n+1}^{p} A_j - \sum_{j=m+1}^{q} B_j > 0$$

the absolutely convergent integral defines a function, analytic for $|\arg(z)| < K\pi/2$ with, in general, singularities at 0 and ∞. If $K = 0$ additional restrictions must be imposed. An extended definition is given by P. Skibiński [113].

From formula (2.9), p.215 of H.M. Srivastava [123] a simple form of the fractional integral for this function can be written

$$(4.2.11) \quad H_{p,q}^{m,n}\left[ax\left|\begin{array}{l}(a_1,1),(a_2,A_2),\cdots,(a_p,A_p)\\(b_1,B_1),\cdots,(b_{q-1},B_{q-1})(b_q,1)\end{array}\right.\right] =$$

$$= I_\phi^{-a_1,a_1-b_q} H_{p-1,q-1}^{m,n-1}\left[ax\left|\begin{array}{l}(a_2,A_2),\cdots,(a_p,A_p)\\(b_1,B_1),\cdots,(b_{q-1},B_{q-1})\end{array}\right.\right],$$

where $n \overset{>}{=} 1$, $q \overset{>}{=} 1$ and $\phi(x) = x$. Hence in the equation

$$(4.2.12) \quad \int_0^x H_{p,q}^{m,n}\left[a(x-t)\left|\begin{array}{l}(a_1,A_1),\cdots,(a_p,A_p)\\(b_1,B_1),\cdots,(b_q,B_q)\end{array}\right.\right]f(t)\ dt = g(x),$$

in the particular case of $A_1 = B_q = 1$ we can use (4.2.9) to factor the operator and consider a new equation

$$(4.2.13) \quad \int_0^x H_{p-1,q-1}^{m,n-1}\left[a(x-t)\left|\begin{array}{l}(a_2,A_2),\cdots,(a_p,A_p)\\(b_1,B_1),\cdots,(b_{q-1},B_{q-1})\end{array}\right.\right]f(t)\ dt = g_1(x)$$

where $g(x) = I_\phi^{-a_1,a_1-b_q} g_1(x)$. In fact, if $A_1 = B_q$ a similar process can be applied; use is made of the fractional integral of Example 1. It should be noted that the H-function reduces to the G-function if all $A_j = B_j = 1$, so that the reduction is quite suitable to G-function kernels. Iteration of the procedure will, of course, reduce the indices further; since $0 \overset{<}{=} m \overset{<}{=} q$ and $0 \overset{<}{=} n \overset{<}{=} p$ only a certain amount of reduction is possible. Since a $G_{p,q+1}^{1,p}$ reduces to a $_pF_q$ function, such kernels can be reduced to $_1F_0$ if $p = q + 1$

or to $_0F_{q-p}$ if $p \overset{<}{=} q$.

(X) Further, in regard to the H-function in [125] Srivastava and Buschman have recently obtained a solution to the equation.

$$(4.2.14) \quad \int_0^x (x-t)^{\rho-1} H_{p,q}^{1,n}\left[x-t \left| \begin{array}{l} (a_1,A_1),\cdots,(a_p,A_p) \\ (0,1),(b_2,B_2),\cdots,(b_q,B_q) \end{array} \right.\right] f(t)\, dt = g(x),$$

with $g^{(\ell)}(0) = 0$ for $0 \overset{<}{=} \ell \overset{<}{=} r-1$, $\mathrm{Re}(\rho) > 0$, and with suitable restrictions on the parameters. Either the method of Mikusiński operators or of Laplace transformations is suitable.

The Laplace transformation of the kernel can be computed directly from the defining integral (4.2.9) and we have

$$\begin{aligned}
(4.2.15) \quad & L\left\{ t^{\rho-1} H_{p,q}^{m,n}\left[t \left| \begin{array}{l} (a_1,A_1),\cdots,(a_p,A_p) \\ (b_1,B_1),\cdots,(b_q,B_q) \end{array} \right.\right]; w \right\} \\
& = w^{-\rho} H_{p+1,q}^{m,n+1}\left[w^{-1} \left| \begin{array}{l} (1-\rho,1),(a_1,A_1),\cdots,(a_p,A_p) \\ (b_1,B_1),\cdots,(b_q,B_q) \end{array} \right.\right],
\end{aligned}$$

provided $\mathrm{Re}(w) > 0$ and $\mathrm{Re}(\rho+b_j/B_j) > 0$, $j = 1, \cdots, m$. Consequently the Laplace transform (in the Mikusiński operator form) of equation (4.2.14) can be written

$$(4.2.16) \quad w^{-\rho} H_{p+1,q}^{m,n+1}\left[w^{-1} \left| \begin{array}{l} (1-\rho,1),(a_1,A_1),\cdots,(a_p,A_p) \\ (b_1,B_1),\cdots,(b_q,B_q) \end{array} \right.\right] f = g.$$

It is desired that the reciprocal of this H-function, or some closely related function, be known, in order to make use of equation (4.2.16).

By formula (1.8), p. 127 of P. Skibiński [113] or formula (6.5), p. 279 of B.L.J. Braaksma [11] we have

$$
H_{p,q}^{m,n}\left[z \left|\begin{array}{c}(a_1,A_1),\cdots,(a_p,A_p)\\(b_1,B_1),\cdots,(b_q,B_q)\end{array}\right.\right]
$$

(4.2.17)

$$
= \sum_{h=1}^{m}\sum_{\nu=0}^{\infty}\frac{\prod\limits_{j=1,j\neq h}^{m}\Gamma\left(b_j-B_j(b_h+\nu)/B_h\right)\prod\limits_{j=1}^{n}\Gamma\left(1-a_j+A_j(b_h+\nu)/B_h\right)}{\prod\limits_{j=m+1}^{q}\Gamma\left(1-b_j+B_j(b_h+\nu)/B_h\right)\prod\limits_{j=n+1}^{p}\Gamma\left(a_j-A_j(b_h+\nu)/B_h\right)}
$$

$$
\cdot\frac{(-1)^{\nu}\,z^{(b_h+\nu)/B_h}}{\nu!\,B_h}
$$

from which, under suitable restrictions, we see the possibility of obtaining a single series expansion for z in the neighborhood of 0, i.e. for large w in equation (4.2.16). If we set $m = 1$, $B_1 = 1$, $b_1 = 0$, for example, we obtain

$$
H_{p+1,q}^{1,n+1}\left[w^{-1}\left|\begin{array}{c}(1-\rho,1),(a_1,A_1),\cdots,(a_p,A_p)\\(0,1),(b_2,B_2),\cdots,(b_q,B_q)\end{array}\right.\right]
$$

(4.2.18)

$$
= \sum_{\nu=0}^{\infty}\frac{\Gamma(\rho+\nu)\prod\limits_{j=1}^{n}\Gamma(1-a_j+A_j\nu)}{\prod\limits_{j=m+1}^{q}\Gamma(1-b_j+B_j\nu)\prod\limits_{j=n+1}^{p}\Gamma(a_j-A_j\nu)}\frac{(-w^{-1})^{\nu}}{\nu!}.
$$

which is an analytic function of $(-w^{-1})$. If we shorten the notation of (4.2.18) to the form

(4.2.19)
$$
H_{p+1,q}^{1,n+1}\left[w^{-1}\left|\begin{array}{c}(1-\rho,1),\text{---}\\(0,1),\text{---}\end{array}\right.\right] = \sum_{\nu=0}^{\infty}c_\nu\,w^{-\nu},
$$

and let k denote the least ν for which $c_\nu \neq 0$, then

(4.2.20)
$$
H_{p+1,q}^{1,n+1}\left[w^{-1}\left|\begin{array}{c}(1-\rho,1),\text{---}\\(0,1),\text{---}\end{array}\right.\right] = w^{-k}\sum_{\mu=0}^{\infty}c_{\mu+k}\,w^{-\mu},
$$

with $c_k \neq 0$, so that the series can be reciprocated. Writing

$$(4.2.21) \qquad \left(\sum_{\mu=0}^{\infty} c_{\mu+k} \, w^{-\mu} \right)^{-1} = \sum_{\lambda=0}^{\infty} C_{\lambda} \, w^{-\lambda},$$

we have convergence for $|w| > \varepsilon$ because the series to be reciprocated has leading coefficient not zero and hence the function has no zeros for $|w^{-1}| < \varepsilon$ for some $\varepsilon > 0$. We assume that the parameters are such that the numbers c_{ν} are all defined; this requires that $\rho + \nu$ and $1 - a_j + A_j \nu$ for $1 \leq j \leq n$ do not take on values which are zero or negative integers for $\nu = 0, 1, 2, \cdots$. The requirement that $c_k \neq 0$ also restricts the parameters in that $1 - b_j + B_j k$ for $m + 1 \leq j \leq q$ and $a_j - A_j k$ for $n + 1 \leq j \leq p$ cannot be zero or negative integers.

Under these restrictions we can now write equation (4.2.16) in the form

$$(4.2.22) \qquad f = \left[w^{-(r-k-\rho)} \sum_{\lambda=0}^{\infty} C_{\lambda} \, w^{-\lambda} \right] \left[w^r g \right],$$

which can be interpreted conveniently and simply as a convolution integral provided $\mathrm{Re}(r-k-\rho) > 0$ and r is an integer (≥ 1).

THEOREM. The convolution equation

$$(4.2.23) \qquad \int_0^x (x-t)^{\rho-1} \, H_{p,q}^{1,n} \left[x-t \, \left| \, \begin{array}{l} (a_1, A_1), \cdots, (a_p, A_p) \\ (0,1), (b_2, B_2), \cdots, (b_q, B_q) \end{array} \right. \right] f(t) \, dt = g(x),$$

with $g^{(\ell)}(0) = 0$ for $0 \leq \ell < r$, $\mathrm{Re}(\rho) > 0$, and with other suitable restrictions on the parameters, has its solution given by

$$f(x) =$$

$$(4.2.24) \qquad = \int_0^x (x-t)^{r-k-\rho-1} \, R_{r-k-\rho} \left[x-t \, \left| \, \begin{array}{l} (a_1, A_1), \cdots, (a_p, A_p) \\ (b_2, B_2), \cdots, (b_q, B_q) \end{array} \right. \right] D_t^r \{ g(t) \} dt,$$

where

$$(4.2.25) \quad R_{r-k-\rho}\left[t \left| \begin{array}{c} (a_1,A_1),\cdots,(a_p,A_p) \\ (b_2,B_2),\cdots,(b_q,B_q) \end{array} \right. \right] = \sum_{\lambda=0}^{\infty} \frac{C_\lambda}{\Gamma(\lambda-k+r-\rho)} \, t^\lambda,$$

and the C_λ can be determined from (4.2.20) and (4.2.21).

From (4.2.22) we obtain (4.2.24) by use of the Laplace transform

$$(4.2.26) \quad L\left\{ t^{r-k-\rho-1} \, R_{r-k-\rho}\left[t \left| \begin{array}{c} (a_1,A_1),\cdots,(a_p,A_p) \\ (b_2,B_2),\cdots,(b_q,B_q) \end{array} \right. \right]; \, w \right\}$$

$$= w^{-(r-k-\rho)} \sum_{\lambda=0}^{\infty} C_\lambda \, w^{-\lambda}.$$

From (4.2.21) we can express the C_λ in terms of the $c_{\mu+k}$ by either the recurrence

$$(4.2.27) \quad C_0 = 1/c_k, \quad \text{and for } \sigma > 0, \quad \sum_{\lambda=0}^{\sigma} C_\lambda \, c_{\sigma+k-\lambda} = 0,$$

or by

$$(4.2.28) \quad C_\lambda = (-1)^\lambda (c_k)^{-\lambda-1} \det \begin{bmatrix} c_{k+1} & c_k & 0 & 0 & \cdots & 0 \\ c_{k+2} & c_{k+1} & c_k & 0 & \cdots & 0 \\ \cdot & & & & & \\ \cdot & & & & & \\ \cdot & & & & & \\ c_{k+\lambda} & c_{k+\lambda-1} & & & \cdots & c_{k+1} \end{bmatrix}$$

From the power series expansion in (4.2.25) it can be seen that the function R is an entire function and the resolvent kernel has the expression

$$K_{r-k-\rho}\left[z \left| \begin{array}{c} \text{—} \\ \text{—} \end{array} \right. \right] = z^{r-k-\rho-1} R_{r-k-\rho}\left[z \left| \begin{array}{c} \text{—} \\ \text{—} \end{array} \right. \right]$$

$$(4.2.29) \quad = \sum_{\lambda=0}^{\infty} C_\lambda \, \frac{z^{\lambda+r-k-\rho-1}}{\Gamma(\lambda+r-k-\rho)},$$

which converges for $|z| > 0$. It may be worth noting that

$$(4.2.30) \qquad D_z K_{r \to k-\rho}\left[z \; \middle| \; \frac{-}{-} \right] = K_{r-k-\rho-1}\left[z \; \middle| \; \frac{-}{-} \right],$$

and that an integral relation involving the H and R functions can be obtained from the direct substitution of (4.2.24) into (4.2.23).

Some special cases of the convolution equation (4.2.23) are as follows:

(I) If we first set $n = p$ in (4.2.23), then we have

$$(4.2.31) \qquad \int_0^x (x-t)^{\rho-1} \; {}_p\Psi_{q-1}\left[1-a_1, \cdots, 1-a_p; 1-b_2, \cdots, 1-b_q; -(x-t) \right]$$
$$\cdot \; f(t) \; dt = g(x),$$

where ${}_p\Psi_q$ denotes Wright's generalized hypergeometric function.

(II) If $A_j = B_j = 1$, then, of course, the equation (4.2.23) becomes

$$(4.2.32) \qquad \int_0^x (x-t)^{\rho-1} \; G_{p,q}^{1,n}\left[x-t \; \middle| \; \begin{matrix} a_1, \cdots, a_p \\ 0, b_2, \cdots, b_q \end{matrix} \right] f(t) \; dt = g(x),$$

in which the kernel involves a special case of the G-function.

(III) If further we set $n = p$ in (4.2.32), we can write

$$\int_0^x (x-t)^{\rho-1} \; {}_pF_{q-1}\left[1-a_1, \cdots, 1-a_p; 1-b_2, \cdots, 1-b_q; -(x-t) \right] f(t) \; dt$$
$$(4.2.34) \qquad = \left(\prod_{j=2}^q \Gamma(1-b_j) \middle/ \prod_{j=1}^p \Gamma(1-a_j) \right) g(x).$$

Other kernels which can be treated in a similar manner are any of those of §4.24 in reference [42], since the Laplace transforms have series representations. In these cases as well, the resolvent kernel is a product of a power function and an entire function for which the series representation can be obtained.

Example (Kummer's function). If $p = 1$, $q = 2$, $1 - a_1 = b$, $1 - b_2 = c$, then (4.2.33) becomes

$$(4.2.34) \quad \int_0^x (x-t)^{\rho-1} \, {}_1F_1 \left[b;c;-(x-t) \right] f(t) \, dt = \left\{ \Gamma(c)/\Gamma(b) \right\} g(x).$$

Since

$$L \left\{ t^{\rho-1} \, {}_1F_1 \left[b;c;-t \right]; w \right\} = L \left\{ t^{\rho-1} e^{-t} \, {}_1F_1 \left[c-b;c;t \right]; w \right\}$$

$$(4.2.35) \qquad\qquad = \Gamma(\rho)(w+1)^{-\rho} \, {}_2F_1 \left[c-b,\rho;c;(w+1)^{-1} \right]$$

$$= \Gamma(\rho) \, w^{-\rho} \, {}_2F_1 \left[b,\rho;c;-w^{-1} \right]$$

for large w, equation (4.2.16) becomes

$$(4.2.36) \quad \left(w^{r-\rho} \left\{ \Gamma(b)\Gamma(\rho)/\Gamma(c) \right\} \, {}_2F_1 \left[b,\rho;c;-w^{-1} \right] \right) f = w^r g.$$

The reciprocal function has coefficients (assuming $k = 0$, $\mathrm{Re}(b) > 0$, $\mathrm{Re}(c) > 0$) of the form

$$(4.2.37) \quad C_\lambda = (-1)^\lambda \det \begin{vmatrix} \dfrac{(b)_1(\rho)_1}{(c)_1\,1!} & 1 & 0 & 0 & \cdots & 0 \\[2ex] \dfrac{(b)_2(\rho)_2}{(c)_2\,2!} & \dfrac{(b)_1(\rho)_1}{(c)_1\,1!} & 1 & 0 & \cdots & 0 \\[1ex] \vdots & & & & & \\[1ex] \dfrac{(b)_\lambda(\rho)_\lambda}{(c)_\lambda\,\lambda!} & & & \cdots & & \dfrac{(b)_1(\rho)_1}{(c)_1\,1!} \end{vmatrix}$$

since $\left\{\Gamma(b)\Gamma(\rho)/\Gamma(c)\right\}\ _2F_1\left[b,\rho;c;0\right] = 1$. Hence

$$(4.2.38) \qquad K_{r-\rho}\left[\ t\ \left|\ \begin{matrix}(1-b,1)\\(1-c,1)\end{matrix}\right.\right] = \sum_{\lambda=0}^{\infty} C_\lambda \frac{t^{\lambda+r-\rho-1}}{\Gamma(\lambda+r-\rho)}\ ,$$

and it is not clear that this can be expressed as a recognizable well-known function.

If $\rho = c$, we have a much simplified C_λ, since

$$(4.2.39) \qquad \Gamma(b)\ _2F_1\left[b,c;c;-w^{-1}\right] = \Gamma(b)\ _1F_0\left[b;-;-w^{-1}\right] = \Gamma(b)\left(1+w^{-1}\right)^{-b},$$

and the reciprocal is

$$(4.2.40) \qquad \left(1+w^{-1}\right)^b\big/\ \Gamma(b).$$

Hence

$$(4.2.41) \qquad f = \left[w^{(r-c)}\left(1+w^{-1}\right)^b\big/\ \Gamma(b)\right](w^r g).$$

Since

$$(4.2.42) \qquad L\left\{\left[t^{c-1}/\Gamma(c)\right]\ _1F_1\left[a;c;-t\right];w\right\} = w^{-c}\ _1F_0\left[a;-;-w^{-1}\right] = w^{-c}\left(1+w^{-1}\right)^a,$$

we have

$$(4.2.43) \qquad f(x) = \left\{\Gamma(r-c)\Gamma(b)\right\}^{-1}\int_0^x (x-t)^{r-c-1}\ _1F_1\left[-a;r-c;-(x-t)\right]D_t^r\\ \left\{g(t)\right\}dt,$$

if $g^{(\ell)}(0) = 0$ for $0 \overset{\leq}{=} \ell < r$. This is a known result [147, p. 44].

4.3 Confluent hypergeometric functions of several variables

The kernel which involves the confluent hypergeometric function of r variables can be treated by means of Mikusinski operators as follows. Consider the integral equation

$$(4.3.1) \quad \int_0^x \frac{(x-t)^{\gamma-1}}{\Gamma(\gamma)} \, e^{\alpha(x-t)} \, \phi_2^r \left[\beta_1, \cdots, \beta_r; \gamma; \lambda_1(x-t), \cdots, \lambda_r(x-t) \right] f(t) dt$$
$$= g(x),$$

where the confluent function ϕ_2^r is defined by

$$\phi_2^r \left[a_1, \cdots, a_r; b; z_1, \cdots, z_r \right]$$

$(4.3.2)$

$$= \frac{\Gamma(b)}{\Gamma(a_1) \cdots \Gamma(a_r)} \sum_{m_1, \cdots, m_r = 0}^{\infty} \frac{\Gamma(a_1 + m_1) \cdots \Gamma(a_r + m_r)}{\Gamma(b + m_1 + \cdots + m_r)} \frac{z_1^{m_1}}{m_1!} \cdots \frac{z_r^{m_r}}{m_r!} .$$

Note, in particular, that

$$\phi_2^r \left[a_1, \cdots, a_r; b; z_1, 0, \cdots, 0 \right] = \phi_2^r \left[a_1, 0, \cdots, 0; b; z_1, \cdots, z_r \right]$$

$(4.3.3)$
$$= {}_1F_1(a_1; b; z_1)$$

and

$$(4.3.4) \quad \lim_{a_1, \cdots, a_r \to \infty} \phi_2^r \left[a_1, \cdots, a_r; b; z_1/a_1, \cdots, z_r/a_r \right] = {}_0F_1(-; b; z_1 + \cdots + z_r)$$

where ${}_1F_1$ and ${}_0F_1$ are the familiar confluent hypergeometric functions connected by Kummer's second formula

$$(4.3.5) \quad e^{-z} \, {}_1F_1(a; 2a; 2z) = {}_0F_1(-; a+1/2; z^2/4).$$

Thus the integral equation (4.3.1), which was solved recently by H.M. Srivastava [124] by using Laplace transforms, provides a generalization of several simpler equations including (1.6.1), (1.8.1), and (3.5.1).

From [42:4.24(5)] we have

$$(4.3.6) \quad L\left\{ \frac{t^{\gamma-1}}{\Gamma(\gamma)} \, \phi_2^r \left[\beta_1, \cdots, \beta_r; \gamma; \lambda_1 t, \cdots, \lambda_r t \right]; s \right\}$$
$$= s^{-\gamma} \left(1 - \lambda_1 s^{-1} \right)^{-\beta_1} \cdots \left(1 - \lambda_r s^{-1} \right)^{-\beta_r},$$

for $\text{Re}(\gamma) > 0$, $\text{Re}(p) > \max_{1 \leq j \leq r} \{0, \lambda_j\}$, and the operator form of

equation (4.3.1) is

$$(4.3.7) \qquad (s-\alpha)^{-\gamma} \left(1 - \lambda_1 (s-\alpha)^{-1}\right)^{\beta_1} \cdots \left(1 - \lambda_r (s-\alpha)^{-1}\right)^{\beta_r} f = g.$$

First consider (4.3.7) rewritten in the form

$$(4.3.8) \qquad (s-\alpha)^{-(\gamma-\beta_1-\cdots-\beta_r)} \left(s - (\alpha+\lambda_1)\right)^{\beta_1} \cdots \left(s - (\alpha+\lambda_r)\right)^{\beta_r} f = g$$

and note that from the correspondence

$$(4.3.9) \qquad L\left\{ t^{\mu-1} e^{\lambda t} / \Gamma(\mu); s \right\} = (s-\lambda)^{-\mu}$$

we obtain the correspondence

$$(4.3.10) \qquad e^{\lambda t} I^{\mu} e^{-\lambda t} \leftrightarrow (s-\lambda)^{-\mu}$$

where I^{μ} is the fractional integral operator of order μ defined

by (2.4.1). Consequently, we can rewrite the original equation in terms

of fractional integral operators and exponential multipliers. Indeed

we have

$$(4.3.11) \qquad \left(e^{\alpha t} I^{\gamma-\beta_1-\cdots-\beta_r} e^{-\alpha t} \right) \left(e^{(\alpha+\lambda_1)t} I^{\beta_1} e^{-(\alpha+\lambda_1)t} \right) \cdots$$
$$\cdots \left(e^{(\alpha+\lambda_r)t} I^{\beta_r} e^{-(\alpha+\lambda_r)t} \right) f(t) = g(t).$$

which simplifies to

$$(4.3.12) \qquad I^{\gamma-\beta_1-\cdots-\beta_r} e^{\lambda_1 t} I^{\beta_1} e^{(\lambda_2-\lambda_1)t} \cdots$$
$$\cdots e^{(\lambda_r-\lambda_{r-1})t} I^{\beta_r} e^{-\lambda_r t} \left(e^{-\alpha t} f(t) \right) = \left(e^{-\alpha t} g(t) \right).$$

The solution can thus be written in the form

$$(4.3.13) \quad \left(e^{-\alpha t}f(t)\right) = e^{\lambda_r t} I^{-\beta_r} e^{(\lambda_{r-1}-\lambda_r)t} \cdots$$

$$\cdots e^{(\lambda_1-\lambda_2)t} I^{-\beta_1} e^{-\lambda_1 t} I^{\beta_1+\cdots+\beta_r-\gamma}\left(e^{-\alpha t}g(t)\right).$$

This format seems to emphasize the fact that the exponential function could be absorbed in the f and g functions. Note that $I^\mu \equiv I^{\mu+n} D^n$, if $\mu \leq 0$, where $n > \mathrm{Re}(\mu)$ is to be understood.

The conditions imposed on g in order that the solution be a function can be read off from the final form. Since in $I^{\beta_1+\cdots+\beta_r-\gamma}$ each of the I^{β_j} operators is compensated for by an $I^{-\beta_j}$ operator, we need $I^{-\gamma}\left(e^{-\alpha t}g(t)\right)$, or merely $I^{-\gamma} g(t)$ to be a function. Thus for $\mathrm{Re}(\gamma) > 0$, if we have $g^{(m)}(0) = 0$ for $0 \leq m < \gamma$ amd $g^{([\gamma]+1)}(t)$ a continuous function, the solution will be continuous. Alternatively, this could be read off from the operator form of the equation if written as

$$(4.3.14) \quad s^{-\gamma}\left(1+Q(s)\right)g = f,$$

where $Q(s) = 0\left(s^{-1}\right)$ as $s \to \infty$, thus we need that $s^\gamma f$ corresponds to a function in order that g corresponds to a function.

Returning to the original operator form of the equation (4.3.7) and solving for f we write

$$(4.3.15) \quad f = (s-\alpha)^{-(n-\gamma)}\left(1-\lambda_1(s-\alpha)^{-1}\right)^{-\beta_1} \cdots \left(1-\lambda_r(s-\alpha)^{-1}\right)^{-\beta_r}\left((s-\alpha)^n g\right)$$

which corresponds to Srivastava's solution [124]

$$(4.3.16) \quad f(x) = \int_0^x \frac{(x-t)^{n-\gamma-1}}{\Gamma(n-\gamma)} e^{\alpha(x-t)}$$

$$\cdot \phi_2^r\left[-\beta_1,\cdots,-\beta_r;n-\gamma;\lambda_1(x-t),\cdots,\lambda_r(x-t)\right]\left((D_t-\alpha)^n g(t)\right)dt,$$

provided $0 < \text{Re}(\gamma) < n$ and $g(x) \in C^n$ for $0 \leq x < \infty$, with $g^{(m)}(0) = 0$, $0 \leq m < n$.

It might be of interest to note that if $\lambda_r = -\alpha$ a minor degeneracy occurs in the solution format since

$$(4.3.17) \qquad (s-\alpha)^{-\gamma}\left[1+\alpha(s-\alpha)^{-1}\right]^{-\beta_r} = (s-\alpha)^{-\gamma-\beta_r} s^{\beta_r}.$$

Now if $0 < \gamma+\beta_r < k$ and $0 < -\beta_r < m$, then we can write

$$f = \left(s-\alpha\right)^{-(k-\gamma-\beta_r)}\left(1-\lambda_1(s-\alpha)^{-1}\right)^{-\beta_1} \cdots$$

$$(4.3.18)$$

$$\cdots \left(1-\lambda_{r-1}(s-\alpha)^{-1}\right)^{-\beta_{r-1}}\left((s-\alpha)^k s^{-m+\beta_r} s^m g\right)$$

and consequently

$$f(x) = \int_0^x \frac{(x-t)^{k-\gamma-\beta_r-1}}{\Gamma(k-\gamma-\beta_r)} e^{-\lambda_r(x-t)}$$

$$(4.3.19)$$

$$\cdot \Phi_2^{r-1}\left[-\beta_1,\cdots,-\beta_{r-1};k-\gamma-\beta_r;\lambda_1(x-t),\cdots,\lambda_{r-1}(x-t)\right]$$

$$\cdot \left(D_t-\lambda_r\right)^k D_t^m I^{m-\beta_r} g(t)\, dt.$$

If $\beta_r \leq 0$, then the expression $D_t^m I^{m-\beta_r} g(t)$ can simply be replaced by $I^{-\beta_r} g(t)$.

4.4 Some open questions

As mentioned in section 3.7 certain methods do not seem to have been explored in the literature beyond those kernels to which they were first applied. The Laplace transformation – Mikusiński operator technique could be more widely used if further progress could be made on the difficult problem, "Given an operator, determine if it corresponds to a function, and if so, to what function."

In particular, in regard to hypergeometric functions it should be noted that the kernels which involve $_1F_1(a;c;u)$ and $_2F_1(a,b;c;u)$ have been solved and the solutions have nice forms if there is a power function multiplier of the form $u^{c-1}/\Gamma(c)$, but not for other powers. For the $_0F_q$ and $_2F_q$ kernels with $q \overset{>}{=} 2$, which arose in the reductions of section 4.2, no nice form is known for the solutions even for $q = 2$ regardless of the choice of the power function multiplier. Some insight for $_pF_q$ and more general kernels might be gained from study of these special cases. However, Prabhakar [86] has pointed out that for $_pF_q$ in general one does not have the factorization of the operator analogous to the $_2F_1$ case which was treated by him and by Love.

Possibly efforts should be concentrated on the H-function of Fox as kernel in the examples of the convolutions of Chapter 2; or at least to the G-function or Wright's $_p\Psi_q$ function kernels. The recent works of Srivastava [123], and Srivastava and Buschman [125], provide a beginning in this direction, some of which has been discussed in Examples IX and X of Section 4.2.

APPENDIX

LIST OF SYMBOLS

arcsin   inverse of sine function

A_n equation (4.1.1)

bei, ber   Kelvin's function [41:7.2.3]

C   Fresnel integral [41:9.10]

Ci   Cosine integral [41:9.8]

C_n^λ Gegenbauer polynomial [41.10.9]

$c_n(x;a)$ Charlier polynomial [41.10.25]

D_ν parabolic cylinder function [41.8.2]

Erf, Erfc, Erfi   error functions [41:6.9.2(23),(24)]

E_α, $E_{\alpha,\beta}$ Mittag-Leffler functions [41:18.1]

$E_{\alpha,\beta}^\rho$ generalized Mittag-Leffler function [83, 84]

Ei, $\overline{\text{Ei}}$ exponential integral [41.6.9.2(25)]

F_1, F_3 Appell's functions [41:5.7.1]

$_1F_1$ Kummer's (confluent hypergeometric) function [41:6]

$_2F_1$ Gauss' (hypergeometric) function [41:2]

$_pF_q$ Generalized hypergeometric function [41:4]

$G_{p,q}^{m,n}$ Meijer's G-function [41:5.3-5.6]

$h_i(x,n)$ hyperbolic functions of higher order [41:18.2]

H_n Hermite polynomial [41:10.13]

H_ν Struve's function [41:7.5.4]

$H_{p,q}^{m,n}$ Fox's H-function [47, 123]

I_ϕ^α, $I_\phi^{\eta,\alpha}$, $I_{x,a}^\alpha$, $I_{\phi,a}^{\eta,\alpha}$ fractional integrals, see section 2.4.

I_ν Modified Bessel function of the first kind [41:7]

J_ν Bessel function of the first kind [41:7]

J_ν^μ Wright's generalized Bessel function, see $\phi(\alpha,\beta,z)$

$k_i(x,n)$ trigonometric functions of higher order [41.18.2]

K^α, $K_\phi^{\eta,\alpha}$, $K_{x,b}^\alpha$, $K_{\phi,b}^{\eta,\alpha}$ fractional integrals, see section 2.4.

K_ν Bessel function of the third kind [41:7]

ℓi logarithmic integral [41:9.7]

$L_n^{(\alpha)}$ Laguerre polynomial [41:10.12]

L Laplace transformation, see equation (3.3.3)

L_T Finite Laplace transformation, see equation (2.3.4)

M Mellin transformation, see equation (3.4.1)

$M_{k,m}$ Whittaker's function [41:6.9]

P_n Legendre polynomial [41:10.10]

P_ν^λ Legendre function [41:10.3]

P_ν^λ Legendre function on the cut [41:10.3.4]

$P_n^{(\alpha,\beta)}$ Jacobi polynomial [41:10.8]

R   generalized hypergeometric function [58, 9]   (Also under G

       and H functions)

S   Fresnel integral [41:9.10]

Si   Sine integral [41.9.8]

$\sinh^{-1}$ inverse of hyperbolic sine function

T   generalized hypergeometric function [58, 9]   (Also under G

       and H functions)

T_n Chebyshev polynomial of the first kind [41:10.11]

T_ϕ Integral transformation, see equation (2.3.2)

U   Heaviside unit function, see section 3.4

U_n Chebyshev polynomial of the second kind [41:10.11]

$W_{\kappa,\mu}$ Whittaker's function [41:6.9]

Γ Gamma function [41:1]

δ Dirac delta function

θ_2, θ_3 Elliptic functions [41:13]

Λ a linear (differential) operator, see section 3.2

$\nu(t)$, $\nu(t,a)$ Volterra's functions [41:18.3]

$\phi(\alpha,\beta,x)$ Wright's function [41:18.1]

Φ_1, Φ_2, Φ_3 confluent hypergeometric functions of two variables

                                               [41:5.7]

Φ_2^r confluent hypergeometric function of r variables

                               [42:Vol.I,p.385; Vol.II,p.445]

$\psi = \Gamma'/\Gamma$ [41:1.7]

Ψ_1 confluent hypergeometric function of two variables [41.5.7]

$_p\Psi_q$ Wright's generalized hypergeometric function, see equation

(4.2.31)

ξ, $\overline{\xi}$ ξ^* generalized hypergeometric functions of Nørlund [76]

(Also under G and H functions)

Ξ_2 confluent hypergeometric function of two variables [41.5.7]

[ ]   greatest integer function in certain formulas of section

4.1, and also in the tables

^   Laplace transform of, see section 2.3

⊙   composition of functions, see section 2.1

*, *$_\phi$ convolution symbols, see section 2.1

INVERSION TABLES

1. Algebraic Functions

2. Exponential, Trigonometric, and Hyperbolic Functions and
   Their Inverses

3. Chebyshev Polynomials

4. Legendre Polynomials and Functions

5. Gegenbauer and Jacobi Polynomials

6. Laguerre and Hermite Polynomials

7. Bessel Functions

8. Confluent Hypergeometric Functions of One and More Variables

9. Hypergeometric Functions of One and More Variables

10. Generalized Hypergeometric Functions

11. G and H Functions

12. Miscellaneous Functions

1. Algebraic Functions

1. $\displaystyle\int_0^x (x-t)^{-1/2} f(t)\, dt = g(x)$

   $g(0) = 0$

   $f(x) = \pi^{-1} \displaystyle\int_0^x (x-t)^{-1/2} g'(t)\, dt$

2. $\displaystyle\int_0^x (x-t)^{1/2} f(t)\, dt = g(x)$

   $g(0) = g'(0) = 0$

   $f(x) = (2/\pi) \displaystyle\int_0^x (x-t)^{-1/2} g''(t)\, dt$

3. $\displaystyle\int_0^x (x-t)^{n-1} f(t)\, dt = g(x)$

   $g^{(r)}(0) = 0, \quad 0 \overset{\le}{=} r < n$

   $f(x) = \big((n-1)!\big)^{-1} g^{(n)}(x)$

4. $\displaystyle\int_0^x (x-t)^{\nu-1} f(t)\, dt = g(x)$

   $g^{(r)}(0) = 0, \quad 0 \overset{\le}{=} r < n, \quad n > \mathrm{Re}(\nu) > 0$

   $f(x) = \big(\Gamma(\nu)\big)^{-1} I^{-\nu} g(x)$

   $\quad = \dfrac{1}{\Gamma(\nu)\Gamma(n-\nu)} \displaystyle\int_0^x (x-t)^{n-\nu-1} g^{(n)}(t)\, dt$

5. $$\int_x^1 (t-x)^{\lambda-1} f(t) \, dt = g(x)$$

$$g^{(r)}(1) = 0, \quad 0 \leq r < n, \quad n > \mathrm{Re}(\lambda) > 0$$

$$f(x) = \frac{(-1)^n}{\Gamma(\lambda)\Gamma(n-\lambda)} \int_x^1 (t-x)^{n-\lambda-1} g^{(n)}(t) \, dt$$

6. $$\int_a^x (x-t)^{-\alpha} f(t) \, dt = g(x)$$

$$g(a) = 0, \quad 0 < \alpha < 1$$

$$f(x) = \frac{\sin\pi\alpha}{\pi} \int_a^x (x-t)^{\alpha-1} g'(t) \, dt$$

7. $$\int_x^b (t-x)^{-\alpha} f(t) \, dt = g(x)$$

$$g(b) = 0, \quad 0 < \alpha < 1$$

$$f(x) = -\frac{\sin\pi\alpha}{\pi} \int_x^b (t-x)^{\alpha-1} g'(t) \, dt$$

8. $$\int_0^x (x^2-t^2)^{-\alpha} f(t) \, dt = g(x)$$

[117]

$$g(0) = 0, \quad 0 < \alpha < 1$$

$$f(x) = \frac{2\sin\pi\alpha}{\pi} \frac{d}{dx} \int_0^x (x^2-t^2)^{\alpha-1} tg(t) \, dt$$

9. $\displaystyle\int_x^\infty (t^2-x^2)^{-\alpha} f(t)\, dt = g(x)$

[117]

$$g(\infty) = 0, \quad 0 < \alpha < 1$$

$$f(x) = -\frac{2\sin\pi\alpha}{\pi} \frac{d}{dx} \int_x^\infty (t^2-x^2)^{\alpha-1}\, tg(t)\, dt$$

10. $\displaystyle\int_0^x \big(1+a(x-t)\big) f(t)\, dt = g(x)$

$$g(0) = g'(0) = 0$$

$$f(x) = \int_0^x e^{-a(x-t)}\, g''(t)\, dt$$

11. $\displaystyle\int_0^x \big((x-t) + a(x-t)^2\big) f(t)\, dt = g(x)$

$$g^{(r)}(0) = 0, \quad 0 \leqq r < 3$$

$$f(x) = \int_0^x e^{-a(x-t)}\, g^{(3)}(t)\, dt$$

12. $\displaystyle\int_1^x t^{-1}(x/t)^{-\alpha} f(t)\, dt = g(x)$

$$g(1) = 0$$

$$f(x) = xg'(x) + \alpha g(x)$$

13. $\underline{FP} \displaystyle\int_0^x (x-t)^{-k} f(t) \, dt = g(x)$

[22]

$k \overset{\geq}{=} 3, \quad g^{(r)}(0) = 0, \quad 0 \overset{\leq}{=} r < k - 1,$

$\xi = \gamma + (-1)^k \displaystyle\sum_{j=0}^{k-2} \frac{(-1)^j}{k-j-1} \binom{k-1}{j}$

$f(x) = \dfrac{(-1)^k \Gamma(k)}{\Gamma(k-2)} \displaystyle\int_0^x (x-u)^{k-3} \int_0^u \nu\big((u-t)e^\xi\big) \, g(t) \, dt \, du$

14. $\underline{FP} \displaystyle\int_0^x (x-t)^{-k-\beta} f(t) \, dt = g(x)$

[22]

$k \overset{\geq}{=} 1, \quad g^{(r)}(0) = 0, \quad 0 \overset{\leq}{=} r < k - 1, \quad \beta \text{ not an integer}$

$f(x) = (-1)^k (k+\beta-1) \dfrac{\sin\pi\beta}{\pi} \displaystyle\int_0^x (x-t)^{k+\beta-2} g(t) \, dt$

2. Exponential, Trigonometric, and Hyperbolic Functions
and Their Inverses

1. $\int_0^x e^{-a(x-t)} f(t)\, dt = g(x)$

   $g(0) = 0$

   $f(x) = g'(x) + ag(x)$

2. $\int_0^x (x-t)^{-1/2} e^{-a(x-t)} f(t)\, dt = g(x)$

   $g(0) = 0$

   $f(x) = \pi^{-1} \int_0^x (x-t)^{-1/2} e^{-a(x-t)} \left(g'(t) + ag(t)\right) dt$

3. $\int_0^x (x-t)^{n-1} e^{-a(x-t)} f(t)\, dt = g(x)$

   $D_x^r g(0) = 0, \quad 0 \overset{\leq}{=} r < n$

   $f(x) = (n-1)!\, (D_x+a)^n g(x)$

4. $\int_0^x (x-t)^{\nu-1} e^{-a(x-t)} f(t)\, dt = g(x)$

   $D_x^r g(0) = 0, \quad 0 \overset{\leq}{=} r < n, \quad n > \mathrm{Re}(\nu) > 0$

   $f(x) = \left(\Gamma(\nu)\right)^{-1} e^{-ax} I^{-\nu} e^{ax} g(x)$

   $= \dfrac{1}{\Gamma(\nu)\Gamma(n-\nu)} \int_0^x (x-t)^{n-\nu-1} e^{-a(x-t)} (D_t+a)^n g(t)\, dt$

5. $\int_0^x \left[1-e^{-a(x-t)}\right] f(t) \, dt = g(x)$

$g(0) = g'(0) = 0$

$f(x) = a^{-1} g''(x) + g'(x)$

6. $\int_0^x \left[1-e^{-a(x-t)}\right]^{n-1} f(t) \, dt = g(x)$

$D_x^r g(0) = 0, \quad 0 \overset{\le}{=} r < n$

$f(x) = \dfrac{a^{1-n}}{(n-1)!} \, D_x(D_x+a) \cdots \left(D_x+(n-1)a\right) g(x)$

7. $\int_0^x \left[1-e^{-a(x-t)}\right]^{\nu-1} f(t) \, dt = g(x)$

$D_x^r g(0) = 0, \quad 0 \overset{\le}{=} r < n, \quad n > \text{Re}(\nu) > 0$

$f(x) = \dfrac{a^{2-n}}{\Gamma(\nu)\Gamma(n-\nu)} \int_0^x e^{-a\nu(x-t)} \left[1-e^{-a(x-t)}\right]^{n-\nu-1} D_t(D_t+a) \cdots \left(D_t+(n-1)a\right) g(t) \, dt$

8. $\int_0^x \left[e^{-a(x-t)} - e^{-b(x-t)}\right] f(t) \, dt = g(x)$

$g(0) = g'(0) = 0$

$f(x) = (b-a)^{-1} \left(g''(x) + (a+b) \, g'(x) + ab \, g(x)\right)$

9. $\displaystyle\int_0^x \sin\big(a(x-t)\big)\, f(t)\, dt = g(x)$

$g(0) = g'(0) = 0$

$f(x) = a^{-1}\, g''(x) + a\, g(x)$

10. $\displaystyle\int_0^x \cos\big(a(x-t)\big)\, f(t)\, dt = g(x)$

$g(0) = 0$

$f(x) = g'(x) + a^2 \displaystyle\int_0^x g(t)\, dt$

11. $\displaystyle\int_0^x \sin^{2n}\big(a(x-t)\big)\, f(t)\, dt = g(x)$

$D_x^r\, g(0) = 0, \quad 0 \overset{<}{=} r < 2n+1$

$f(x) = \dfrac{a^{-2n}}{(2n)!}\, D_x \left[D_x^2 + (2a)^2\right] \cdots \left[D_x^2 + (2an)^2\right] g(x)$

12. $\displaystyle\int_0^x \sin^{2n+1}\big(a(x-t)\big)\, f(t)\, dt = g(x)$

$D_x^r\, g(0) = 0, \quad 0 \overset{<}{=} r < 2n+2$

$f(x) = \dfrac{a^{-2n-1}}{(2n+1)!} \left[D_x^2 + a^2\right]\left[D_x^2 + (3a)^2\right] \cdots \left[D_x^2 + \big((2n+1)a\big)^2\right] g(x)$

13. $\displaystyle\int_0^x (x-t)^{-1/2} \sin\bigl(a(x-t)\bigr)\, f(t)\, dt = g(x)$

$g(0) = g'(0) = 0$

$f(x) = \dfrac{2}{\pi a}\displaystyle\int_0^x (x-t)^{-1/2} \cos\bigl(a(x-t)\bigr)\bigl(g''(t) + a^2 g(t)\bigr)\, dt$

14. $\displaystyle\int_0^x (x-t)^{-1/2} \cos\bigl(a(x-t)\bigr)\, f(t)\, dt = g(x)$

$g(0) = g'(0) = 0$

$f(x) = \dfrac{2}{\pi a}\displaystyle\int_0^x (x-t)^{-1/2} \sin\bigl(a(x-t)\bigr)\bigl(g''(t) + a^2 g(t)\bigr)\, dt$

15. $\displaystyle\int_0^x \sin\bigl(k(x-t)^{1/2}\bigr)\, f(t)\, dt = g(x)$

$g(0) = g'(0) = 0$

$f(x) = \dfrac{4}{\pi k^2}\displaystyle\int_0^x \cosh\bigl(k(x-t)^{1/2}\bigr)\, g''(t)\, dt$

16. $\displaystyle\int_0^x (x-t)^{1/2} \cos\bigl(k(x-t)^{1/2}\bigr)\, f(t)\, dt = g(x)$

$g(0) = 0$

$f(x) = \dfrac{1}{\pi}\displaystyle\int_0^x (x-t)^{-1/2} \cosh\bigl(k(x-t)^{1/2}\bigr)\, g'(t)\, dt$

17. $\int_0^x (\cos t - \cos x)^{-1} f(t)\, dt = g(x)$

[117]

$g(0) = 0, \quad 0 < x < \pi$

$f(x) = \dfrac{1}{\pi} \dfrac{d}{dx} \int_0^x (\cos t - \cos x)^{-1} \sin t\, g(t)\, dt$

18. $\int_0^x \sin\big(a(x-t)\big) \sin\big(b(x-t)\big) f(t)\, dt = g(x)$

$D_x^r\, g(0) = 0, \quad 0 \overset{\le}{=} r < 3$

$f(x) = \left[D_x^2 + (a+b)^2 \right]\left[D_x^2 + (a-b)^2 \right] \dfrac{1}{2ab} \int_0^x g(t)\, dt$

19. $\int_0^x \cos\big(a(x-t)\big) \sin\big(b(x-t)\big) f(t)\, dt = g(x)$

$D_x^r\, g(0) = 0, \quad 0 \overset{\le}{=} r < 3, \quad a^2 > b^2$

$f(x) = \left[D_x^2 + (a+b)^2 \right]\left[D_x^2 + (a-b)^2 \right] \dfrac{1}{b(a^2-b^2)^{1/2}} \int_0^x \sinh\left[(a^2-b^2)^{1/2}(x-t) \right] g(t)\, dt$

20. $\int_0^x \cos\big(a(x-t)\big) \sin\big(b(x-t)\big) f(t)\, dt = g(x)$

$D_x^r\, g(0) = 0, \quad 0 \overset{\le}{=} r < 3, \quad a^2 < b^2$

$f(x) = \left[D_x^2 + (a+b)^2 \right]\left[D_x^2 + (a-b)^2 \right] \dfrac{1}{b(b^2-a^2)^{1/2}} \int_0^x \sin\left[(b^2-a^2)^{1/2}(x-t) \right] g(t)\, dt$

21. $\displaystyle\int_0^x \cos\bigl(a(x-t)\bigr) \cos\bigl(b(x-t)\bigr) f(t)\ dt = g(x)$

$$D_x^r\ g(0) = 0, \quad 0 \overset{\le}{=} r < 3$$

$$f(x) = \left[D_x^2+(a+b)^2\right]\left[D_x^2+(a-b)^2\right](a^2+b^2)^{-1/2} \int_0^x \int_0^u \sin\bigl((a^2+b^2)^{1/2}(u-t)\bigr)\ g(t)\ dt\ du$$

22. $\displaystyle\int_0^x \left[\cos\bigl(b(x-t)\bigr)-\cos\bigl(a(x-t)\bigr)\right] f(t)\ dt = g(x)$

$$D_x^r\ g(0) = 0, \quad 0 \overset{\le}{=} r < 3$$

$$f(x) = \left[D_x^2+a^2\right]\left[D_x^2+b^2\right]\frac{1}{a^2-b^2}\int_0^x g(t)\ dt$$

23. $\displaystyle\int_0^x \sin\bigl(a(x-t)+b\bigr) f(t)\ dt = g(x)$

$$g(0) = g'(0) = 0, \quad b \ne k\pi$$

$$f(x) = \csc b \int_0^x e^{-(a\ \text{ctn}\ b)(x-t)}\bigl(g''(t) + a^2 g(t)\bigr)\ dt$$

24. $\displaystyle\int_0^x \sinh\bigl(a(x-t)\bigr) f(t)\ dt = g(x)$

$$g(0) = g'(0) = 0$$

$$f(x) = a^{-1}g''(x) - ag(x)$$

25. $\int_0^x \cosh\bigl(a(x-t)\bigr)\ f(t)\ dt = g(x)$

$g(0) = 0$

$f(x) = g'(x) - a^2 \int_0^x g(t)\ dt$

26. $\int_0^x \sinh^n\bigl(a(x-t)\bigr)\ f(t)\ dt = g(x)$

$D_x^r g(0) = 0, \quad 0 \leqq r < n{+}1$

$f(x) = \dfrac{a^{-n}}{n!}\ \bigl(D_x{+}na\bigr)\bigl(D_x{+}(n{-}2)a\bigr)\ \cdots\ \bigl(D_x{-}na\bigr)\ g(x)$

27. $\int_0^x (x-t)^{-1/2} \sinh\bigl(a(x-t)\bigr)\ f(t)\ dt = g(x)$

$g(0) = g'(0) = 0$

$f(x) = \dfrac{2}{\pi a} \int_0^x (x-t)^{-1/2} \cosh\bigl(a(x-t)\bigr)\bigl(g''(t){-}a^2 g(t)\bigr)\ dt$

28. $\int_0^x (x-t)^{-1/2} \cosh\bigl(a(x-t)\bigr)\ f(t)\ dt = g(x)$

$g(0) = g'(0) = 0$

$f(x) = \dfrac{2}{\pi a} \int_0^x (x-t)^{-1/2} \sinh\bigl(a(x-t)\bigr)\bigl(g''(t){-}a^2 g(t)\bigr)\ dt$

29. $\displaystyle\int_0^x \sinh\left(k(x-t)^{1/2}\right) f(t) \, dt = g(x)$

$g(0) = g'(0) = 0$

$\displaystyle f(x) = \frac{4}{\pi k^2} \int_0^x \cos\left(k(x-t)^{1/2}\right) g''(t) \, dt$

30. $\displaystyle\int_0^x (x-t)^{-1/2} \cosh\left(k(x-t)^{1/2}\right) f(t) \, dt = g(x)$

$g(0) = 0$

$\displaystyle f(x) = \frac{1}{\pi} \int_0^x (x-t)^{-1/2} \cos\left(k(x-t)^{1/2}\right) g'(t) \, dt$

31. $\displaystyle\int_0^x \left[\cosh\left(a(x-t)\right) - \cosh\left(b(x-t)\right)\right] f(t) \, dt = g(x)$

$D_x^r g(0) = 0, \quad 0 \overset{\leq}{=} r < 3$

$\displaystyle f(x) = \left(D_x^2 - a^2\right)\left(D_x^2 - b^2\right) \frac{1}{a^2 - b^2} \int_0^x g(t) \, dt$

32. $\displaystyle\int_0^x \left[\sinh\left(a(x-t)\right) - \sin\left(a(x-t)\right)\right] f(t) \, dt = g(x)$

$D_x^r g(0) = 0, \quad 0 \overset{\leq}{=} r < 4$

$\displaystyle f(x) = \frac{1}{2a^3} \left(D_x^4 - a^4\right) g(x)$

33. $\int_0^x \sinh\big(a(x-t)\big) \sin\big(a(x-t)\big) f(t) \ dt = g(x)$

$D_x^r \ g(0) = 0, \quad 0 \overset{\le}{=} r < 3$

$f(x) = \left[D_x^4 + 4a^4\right] \dfrac{1}{2a^2} \int_0^x g(t) \ dt$

34. $\int_0^x \sinh\big(a(x-t)\big) \cos\big(a(x-t)\big) f(t) \ dt = g(x)$

$D_x^r \ g(0) = 0, \quad 0 \overset{\le}{=} r < 4$

$f(x) = \left[D_x^4 + 4a^4\right] \dfrac{1}{a^2 \sqrt{2}} \int_0^x \sinh\big(a\sqrt{2}(x-t)\big) \ g(t) \ dt$

35. $\int_0^x \cosh\big(a(x-t)\big) \cos\big(a(x-t)\big) f(t) \ dt = g(x)$

$g(0) = 0$

$f(x) = \left[D_x^4 + 4a^4\right] \dfrac{1}{2} \int_0^x (x-t)^2 \ g(t) \ dt$

36. $\int_0^x \cosh\big(a(x-t)\big) \sin\big(a(x-t)\big) f(t) \ dt = g(x)$

$D_x^r \ g(0) = 0, \quad 0 \overset{\le}{=} r < 4$

$f(x) = \left[D_x^4 + 4a^4\right] \dfrac{1}{a^2 \sqrt{2}} \int_0^x \sin\big(a\sqrt{2}(x-t)\big) \ g(t) \ dt$

37. $\displaystyle\int_0^x \log(x-t)\, f(t)\, dt = g(x)$

$g(0) = g'(0), \quad \gamma = -\psi(1)$

$f(x) = -\displaystyle\int_0^x \nu\big((x-t)/\gamma\big)\, g''(t)\, dt$

38. $\displaystyle\int_0^x (x-t)^{-1/2} \log(x-t)\, f(t)\, dt = g(x)$

$g(0) = 0, \quad \gamma = -\psi(1)$

$f(x) = -\dfrac{1}{2(\pi\gamma)^{1/2}} \displaystyle\int_0^x \nu\big((x-t)/(4\gamma), -1/2\big)\, g'(t)\, dt$

39. $\displaystyle\int_0^x (x-t)^{\alpha-1} \big(\psi(\alpha)-\log(x-t)\big)\, f(t)\, dt = g(x)$

$g^{(r)}(0) = 0, \quad 0 \leq r < n, \quad n > \text{Re}(\alpha) > 0$

$f(x) = \Gamma(\alpha) \displaystyle\int_0^x \nu(x-t,\, n-\alpha-1)\, g^{(n)}(t)\, dt$

40. $\displaystyle\int_0^x \log(x/t)\, f(t)\, dt = g(x)$

$g(0) = g'(0) = 0$

$f(x) = \dfrac{d}{dx}\big(xg'(x)\big)$

41. $\displaystyle\int_0^x \log(t/x)\ f(t)\ dt = g(x)$

   $g(0) = g'(0) = 0$

$f(x) = -\dfrac{d}{dx}\left(xg'(x)\right)$

42. $\displaystyle\int_x^\infty \log(x/t)\ f(t)\ dt = g(x)$

   $g(\infty) = g'(\infty) = 0$

$f(x) = -\dfrac{d}{dx}\left(xg'(x)\right)$

43. $\displaystyle\int_x^\infty \log(t/x)\ f(t)\ dt = g(x)$

   $g(\infty) = g'(\infty) = 0$

$f(x) = \dfrac{d}{dx}\left(xg'(x)\right)$

44. $\displaystyle\int_0^x \log\left[\dfrac{\sqrt{x}+\sqrt{(x-t)}}{\sqrt{x}-\sqrt{(x-t)}}\right]\ f(t)\ dt = g(x)$

   $g(0) = 0$

$f(x) = \pi^{-1/2}\ I_x^{-1/2}\ x^{1/2}\ D_x\ g(x) = \pi^{-1/2}\ D_x\ x\ I_x^{-1/2}\ x^{-1/2}\ g(x)$

45. $\int_x^\infty \log\left(\frac{\sqrt{t}+\sqrt{(t-x)}}{\sqrt{t}-\sqrt{(t-x)}}\right) f(t)\ dt = g(x)$

   $g(0) = 0$

$f(x) = -\pi^{-1/2}\ D\ x^{1/2}\ K^{-1/2}\ g(x) = -\pi^{-1/2}\ x^{-1/2}\ K^{-1/2}\ x\ D\ g(x)$

46. $\int_0^x \arcsin\left((1-t/x)^{1/2}\right) f(t)\ dt = g(x)$

   $g(0) = 0$

$f(x) = 2\pi^{-1/2}\ D_x\ x^{1/2}\ I_x^{-1/2}\ g(x) = 2\pi^{-1/2}\ x^{-1/2}\ i_x^{-1/2}\ x\ D_x\ g(x)$

47. $\int_x^\infty \arcsin\left((1-x/t)^{1/2}\right) f(t)\ dt = g(x)$

   $g(\infty) = 0$

$f(x) = -2\pi^{-1/2}\ K_x^{-1/2}\ x^{1/2}\ D_x\ g(x) = -2\pi^{-1/2}\ D_x\ x\ K_x^{-1/2}\ x^{-1/2}\ g(x)$

48. $\int_0^x \arctan\left((x/t-1)^{1/2}\right) f(t)\ dt = g(x)$

   $g(0) = 0$

$f(x) = 2\pi^{-1/2}\ D_x\ x^{1/2}\ I_x^{-1/2}\ g(x) = 2\pi^{-1/2}\ x^{-1/2}\ I_x^{-1/2}\ x\ D_x\ g(x)$

49. $$\int_x^\infty \arctan\left((t/x-1)^{1/2}\right) f(t)\ dt = g(x)$$

$$g(\infty) = 0$$

$$f(x) = -2\pi^{-1/2}\ D_x\ x^{1/2}\ K_x^{-1/2}\ x^{-1/2}\ g(x) = -2\pi^{-1/2}\ K_x^{-1/2}\ x^{1/2}\ D_x\ g(x)$$

3. Chebyshev Polynomials

1. $$\int_x^1 \left(t^2-x^2\right)^{-1/2} T_n(t/x)\, f(t)\, dt = g(x)$$

[64] $g(1) = 0, \quad 0 < x < 1$

$$f(x) = -\frac{2}{\pi} \int_x^1 \left(t^2-x^2\right)^{-1/2} T_{n-1}(x/t)\, t^{1-n}\, D_t\left(t^n g(t)\right)\, dt$$

2. $$\int_0^x \left(x^2-t^2\right)^{-1/2} T_n(x/t)\, f(t)\, dt = g(x)$$

$g(0) = 0$

$$f(x) = 2\pi^{-1/2}\, x^{n+1}\, I_{x^2}^{-n/2-1/2}\, x^{1-n}\, I_{x^2}^{n/2}\, x^{-1}\, g(x)$$

3. $$\int_0^x \left(x^2-t^2\right)^{-1/2} T_n(t/x)\, f(t)\, dt = g(x)$$

$g(0) = 0$

$$f(x) = 2\pi^{-1/2}\, x\, I_{x^2}^{-n/2-1/2}\, x^n\, I_{x^2}^{n/2}\, x^{-n}\, g(x)$$

4. $$\int_x^b \left(t^2-x^2\right)^{-1/2} T_n(x/t)\, f(t)\, dt = g(x)$$

$g(b) = 0, \quad x > 0, \quad b \stackrel{<}{=} \infty$

$$f(x) = 2\pi^{-1/2}\, x^{n+1}\, K_{x^2,b}^{-n/2-1/2}\, x^{1-n}\, K_{x^2,b}^{n/2}\, x^{-1}\, g(x)$$

5. $\int_{x}^{b} \left(t^2 - x^2\right)^{-1/2} T_n(t/x) \, f(t) \, dt = g(x)$

     $g(b) = 0, \quad x > 0, \quad b \stackrel{\leq}{=} \infty$

$f(x) = 2\pi^{-1/2} \, x \, K_{x^2,b}^{-n/2-1/2} \, x^n \, K_{x^2,b}^{n/2} \, x^{-n} \, g(x)$

6. $\int_{0}^{x} (x-t)^{-1/2} \, T_n(2x/t - 1) \, f(t) \, dt = g(x)$

     $g(0) = 0$

$f(x) = \pi^{-1/2} \, x^n \, D_x^n \, x^{-n} \, I_x^{n-1/2} \, g(x)$

7. $\int_{0}^{x} (x-t)^{1/2} \, T_n(2t/x - 1) \, f(t) \, dt = g(x)$

     $g(0) = 0$

$f(x) = \pi^{-1/2} \, x^{-1/2} \, D_x^n \, x^{n+1/2} \, I_x^{n-1/2} \, x^{-n} \, g(x)$

8. $\int_{x}^{b} (t-x)^{-1/2} \, T_n(2x/t - 1) \, f(t) \, dt = g(x)$

     $g(b) = 0, \quad x > 0, \quad b \stackrel{\leq}{=} \infty$

$f(x) = \pi^{-1/2} \, x^n \, (-D_x)^n \, x^{-n} \, K_{x,b}^{n-1/2} \, g(x)$

9. $\quad \displaystyle\int_x^b (t-x)^{-1/2} \, T_n(2t/x-1) \, f(t) \, dt = g(x)$

$\quad g(b) = 0, \quad x > 0, \quad b \overset{<}{=} \infty$

$f(x) = \pi^{-1/2} \, x^{-1/2} \, (-D_x)^n \, x^{n+1/2} \, K_{x,b}^{n-1/2} \, x^{-n} \, g(x)$

10. $\quad \displaystyle\int_0^x \left(x^2-t^2\right)^{1/2} \, U_n(x/t) \, f(t) \, dt = g(x)$

$\quad g(0) = D_x \, g(0) = 0$

$f(x) = \dfrac{4\pi^{-1/2}}{n+1} \, x^{n+1} \, I_{x^2}^{-n/2-3/2} \, x^{1-n} \, I_{x^2}^{n/2} \, x^{-1} \, g(x)$

11. $\quad \displaystyle\int_0^x \left(x^2-t^2\right)^{1/2} \, U_n(t/x) \, f(t) \, dt = g(x)$

$\quad g(0) = D_x \, g(0) = 0$

$f(x) = \dfrac{4\pi^{-1/2}}{n+1} \, x \, I_{x^2}^{-n/2-3/2} \, x^{n+2} \, I_{x^2}^{n/2} \, x^{-n-2} \, g(x)$

12. $\quad \displaystyle\int_x^b \left(t^2-x^2\right)^{1/2} \, U_n(x/t) \, f(t) \, dt = g(x)$

$\quad g(b) = D_x \, g(b) = 0, \quad x > 0, \quad b \overset{<}{=} \infty$

$f(x) = \dfrac{4\pi^{-1/2}}{n+1} \, x^{n+1} \, K_{x^2,b}^{-n/2-3/2} \, x^{1-n} \, K_{x^2,b}^{n/2} \, x^{-1} \, g(x)$

13. $\displaystyle\int_x^b \left(t^2-x^2\right)^{1/2} U_n(t/x)\, f(t)\, dt = g(x)$

$g(b) = D_x\, g(b) = 0, \quad x > 0, \quad b \overset{<}{=} \infty$

$f(x) = \dfrac{4\pi^{-1/2}}{n+1}\, x\, K_{x^2,b}^{-n/2-3/2}\, x^{n+2}\, K_{x^2,b}^{n/2}\, x^{-n-2}\, g(x)$

14. $\displaystyle\int_0^x (x-t)^{1/2} U_n(2x/t-1)\, f(t)\, dt = g(x)$

$g(0) = D_x\, g(0) = 0$

$f(x) = \dfrac{2\pi^{-1/2}}{n+1}\, x^n\, D_x^{n+2}\, I_x^{n+1/2}\, g(x)$

15. $\displaystyle\int_0^x (x-t)^{1/2} U_n(2t/x-1)\, f(t)\, dt = g(x)$

$g(0) = D_x g(0) = 0$

$f(x) = \dfrac{2\pi^{-1/2}}{n+1}\, x^{3/2}\, D_x^{n+2}\, x^{n+3/2}\, I_x^{n+1/2}\, x^{-n-2}\, g(x)$

16. $\displaystyle\int_x^b (t-x)^{1/2} U_n(2x/t-1)\, f(t)\, dt = g(x)$

$g(b) = D_x\, g(b) = 0, \quad x > 0, \quad b \overset{<}{=} \infty$

$f(x) = \dfrac{2\pi^{-1/2}}{n+1}\, x^{3/2}\, (-D_x)^{n+2}\, x^{n+3/2}\, K_{x,b}^{n+1/2}\, g(x)$

17. $\displaystyle\int_x^b (t-x)^{1/2} U_n(2t/x-1) f(t) \, dt = g(x)$

$g(b) = D_x \, g(b) = 0, \quad x > 0, \quad b \leqq \infty$

$f(x) = \dfrac{2\pi^{-1/2}}{n+1} \, x^{3/2} \left(-D_x\right)^{n+2} x^{n+3/2} \, K_{x,b}^{n+1/2} \, x^{-n-2} \, g(x)$

4. Legendre Polynomials and Functions

1. $$\int_x^1 P_n(x/t) \; f(t) \; dt = g(x)$$

[58]

$$g(1) = D_x \, g(1) = 0$$

$$f(x) = x^{-2} \int_x^1 t^{2+n} \; P_{n-2}(t/x) \left(t^{-1}D_t\right)^2 \left(t^{-n+2}g(t)\right) \, dt$$

2. $$\int_x^1 P_n(t/x) \; f(t) \; dt = g(x)$$

$$g(1) = D_x \, g(1) = 0$$

$$f(x) = \int_x^1 t^{2-n} \; P_{n-2}(x/t) \left(t^{-1}D_t\right)^2 \left(t^n g(t)\right) \, dt$$

3. $$\int_1^x P_n(x/t) \; f(t) \; dt = g(x)$$

[38]

$$g(1) = 0$$

$$f(x) = \frac{x^{n+1}}{(n-1)!} \left(x^{-1}D_x\right)^{n+1} \int_1^x (x-t)^{n-1} \; g(t) \, dt$$

4. $$\int_1^x P_n(t/x) \; f(t) \; dt = g(x)$$

[58]

$$g(1) = D_x \, g(1) = 0$$

$$f(x) = \int_1^x t^{2-n} \; P_{n-2}(x/t) \left(t^{-1}D_t\right)^2 \left(t^n g(t)\right) \, dt$$

5. $\quad \displaystyle\int_0^x P_n(x/t) \, f(t) \, dt = g(x)$

$\quad\quad g(0) = D_x \, g(0) = 0$

$f(x) = 2x^{n+1} \, I_{x^2}^{-n/2-1} \, x^{1-n} \, I_{x^2}^{n/2} \, x^{-1} \, g(x)$

6. $\quad \displaystyle\int_0^x P_n(t/x) \, f(t) \, dt = g(x)$

$\quad\quad g(0) = D_x \, g(0) = 0$

$f(x) = 2x \, I_{x^2}^{-n/2-1} \, x^{1+n} \, I_{x^2}^{n/2} \, x^{-n-1} \, g(x)$

7. $\quad \displaystyle\int_x^b P_n(x/t) \, f(t) \, dt = g(x)$

$\quad\quad g(b) = D_x \, g(b) = 0, \quad x > 0, \quad b \stackrel{<}{=} \infty$

$f(x) = 2x^{n+1} \, K_{x^2,b}^{-n/2-1} \, x^{1-n} \, K_{x^2,b}^{n/2} \, x^{-1} \, g(x)$

8. $\quad \displaystyle\int_x^b P_n(t/x) \, f(t) \, dt = g(x)$

$\quad\quad g(b) = D_x \, g(b) = 0, \quad x > 0, \quad b \stackrel{<}{=} \infty$

$f(x) = 2x \, K_{x^2}^{-n/2-1} \, x^{1+n} \, K_{x^2,b}^{n/2} \, x^{-n-1} \, g(x)$

9. $\displaystyle\int_0^x P_n(2x/t-1)\ f(t)\ dt = g(x)$

$g(0) = 0$

$f(x) = x^n\ D_x^{n+1}\ x^{-n}\ I_x^n\ g(x)$

10. $\displaystyle\int_0^x P_n(2t/x-1)\ f(t)\ dt = g(x)$

$g(0) = 0$

$f(x) = D_x^{n+1}\ x^{n+1}\ I^n\ x^{-n-1}\ g(x)$

11. $\displaystyle\int_x^b P_n(2x/t-1)\ f(t)\ dt = g(x)$

$g(b) = 0,\quad x > 0,\quad b \overset{<}{=} \infty$

$f(x) = x^n\ \left(-D_x\right)^{n+1}\ x^{-n}\ K_{x,b}^n\ g(x)$

12. $\displaystyle\int_x^b P_n(2t/x-1)\ f(t)\ dt = g(x)$

$g(b) = 0,\quad x > 0,\quad b \overset{<}{=} \infty$

$f(x) = \left(-D_x\right)^{n+1}\ x^{n+1}\ K_{x,b}^n\ x^{-n-1}\ g(x)$

13.
$$\int_0^x P_n\left(e^{x-t}\right) f(t) \, dt = g(x)$$

[139]

$$g(0) = D_x \, g(x) = 0$$

$$f(x) = \int_0^x P_{n-2}\left(e^{t-x}\right) e^{t-x} e^{(n-2)t} D_t \, e^{2t} D_t \, e^{-nt} g(t) \, dt$$

14.
$$\int_0^x P_n\left(e^{-(x-t)}\right) f(t) \, dt = g(x)$$

$$g(0) = 0$$

$$f(x) = \left(D_x+n\right)\left(D_x+n-2\right) \cdots \left(D_x-n+2\right)\left(e^x I_x\right)^{n-1} e^{(1-n)x} g(x)$$

15.
$$\int_0^x P_{2n}\left(2e^{-(x-t)}-1\right) f(t) \, dt = g(x)$$

$$g(0) = 0$$

$$f(x) = D_x\left(D_x+1\right) \cdots \left(D_x+n\right)\left(e^x I_x\right)^n e^{-nx} g(x)$$

16.
$$\int_0^x P_n\left(\cos(x-t)\right) f(t) \, dt = g(x)$$

[80]

$$g(0) = D_x \, g(0) = 0$$

$$f(x) = \left[D_x^2 + (n+1)^2\right] \int_0^x P_{n+1}\left(\cos(x-t)\right) g(t) \, dt$$

17. $\displaystyle\int_0^x P_n\bigl(\cosh(x-t)\bigr)\, f(t)\, dt = g(x)$

$g(0) = D_x\, g(0) = 0$

$f(x) = \left[D_x^2 - (n+1)^2\right]\displaystyle\int_0^x P_{n+1}\bigl(\cosh(x-t)\bigr)\, g(t)\, dt$

18. $\displaystyle\int_0^x \left(x^2-t^2\right)^{-\mu/2} P_\nu^\mu(x/t)\, f(t)\, dt = g(x)$

$n > \mathrm{Re}(1-\mu) > 0,\quad \mathrm{Re}(\nu) > -1,\quad D_x^r\, g(0) = 0,\quad 0 \stackrel{\leq}{=} r < n$

$f(x) = 2^{1-\mu}\, x^{1+\nu}\, I_{x^2}^{(\mu-\nu-2)/2}\, x^{1-\mu-\nu}\, I_{x^2}^{(\mu+\nu)/2}\, x^{-1}\, g(x)$

19. $\displaystyle\int_0^x \left(x^2-t^2\right)^{-\mu/2} P_\nu^\mu(t/x)\, f(t)\, dt = g(x)$

$n > \mathrm{Re}(1-\mu) > 0,\quad \mathrm{Re}(\nu) > -1,\quad D_x^r\, g(0) = 0,\quad 0 \stackrel{\leq}{=} r < n$

$f(x) = 2^{1-\mu}\, x\, I_{x^2}^{(\mu-\nu-2)/2}\, x^{1-\mu+\nu}\, I_{x^2}^{(\mu+\nu)/2}\, x^{-\nu-1}\, g(x)$

20. $\displaystyle\int_x^b \left(t^2-x^2\right)^{-\mu/2} P_\nu^\mu(x/t)\, f(t)\, dt = g(x)$

$n > \mathrm{Re}(1-\mu) > 0,\quad \mathrm{Re}(\nu) > -1,\quad D_x^r\, g(b) = 0,\quad 0 \stackrel{\leq}{=} r < n,\quad x > 0,$
$\qquad\qquad\qquad\qquad\qquad\qquad\qquad\qquad\qquad\qquad\qquad b \stackrel{\leq}{=} \infty$

$f(x) = 2^{1-\mu}\, x^{1+\nu}\, K_{x^2,b}^{(\mu-\nu-2)/2}\, x^{1-\mu-\nu}\, K_{x^2,b}^{(\mu+\nu)/2}\, x^{-1}\, g(x)$

21. $\displaystyle\int_x^b \left(t^2-x^2\right)^{-\mu/2} P_\nu^\mu(t/x) \, f(t) \, dt = g(x)$

      $n > \mathrm{Re}(1-\mu) > 0, \quad \mathrm{Re}(\nu) > -1, \quad D_x^r \, g(b) = 0, \quad 0 \overset{<}{=} r < n, \quad x > 0,$
                                                                      $b \overset{<}{=} \infty$

$f(x) = 2^{1-\mu} \, x \, K_{x^2,b}^{(\mu-\nu-2)/2} \, x^{1-\mu+\nu} \, K_{x^2,b}^{(\mu+\nu)/2} \, x^{-\nu-1} \, g(x)$

22. $\displaystyle\int_0^x (x-t)^{-\mu/2} P_\nu^\mu(2x/t-1) \, f(t) \, dt = g(x)$

      $n > \mathrm{Re}(1-\mu) > 0, \quad \mathrm{Re}(\nu) > -1, \quad D_x^r \, g(0) = 0, \quad 0 \overset{<}{=} r < n$

$f(x) = x^\nu \, I^{-\nu-1} \, x^{-\nu} \, I^{\nu+\mu} \, x^{-\mu/2} \, g(x)$

23. $\displaystyle\int_0^x (x-t)^{-\mu/2} P_\nu^\mu(2t/x-1) \, f(t) \, dt = g(x)$

      $n > \mathrm{Re}(1-\mu) > 0, \quad \mathrm{Re}(\nu) > -1, \quad D_x^r \, g(0) = 0, \quad 0 \overset{<}{=} r < n$

$f(x) = (-x)^{\mu/2} \, I^{-\nu-1} \, x^{1-\mu+\nu} \, I^{\nu+\mu} \, x^{-\nu-1} \, g(x)$

24. $\displaystyle\int_x^b (t-x)^{-\mu/2} P_\nu^\mu(2x/t-1) \, f(t) \, dt = g(x)$

      $n > \mathrm{Re}(1-\mu) > 0, \quad \mathrm{Re}(\nu) > -1, \quad D_x^r \, g(b) = 0, \quad 0 \overset{<}{=} r < n, \quad x > 0,$
                                                                       $b \overset{<}{=} \infty$

$f(x) = x^\nu \, K_{x,b}^{-\nu-1} \, x^{-\nu} \, K_{x,b}^{\nu+\mu} \, x^{-\mu/2} \, g(x)$

25. $\displaystyle\int_x^b (t-x)^{-\mu/2} \, P_\nu^\mu(2t/x-1) \, f(t) \, dt = g(x)$

$n > \mathrm{Re}(1-\mu) > 0, \quad \mathrm{Re}(\nu) > -1, \quad D_x^r \, g(b) = 0, \quad 0 \leqq r < n, \quad x > 0, \quad b \leqq \infty$

$f(x) = x^{\mu/2} \, K_{x,b}^{-\nu-1} \, x^{1-\mu+\nu} \, K_{x,b}^{\nu+\mu} \, x^{-\nu-1} \, g(x)$

26. $\displaystyle\int_a^x \left(x^2-t^2\right)^{\lambda/2} \, P_\nu^{-\lambda}(x/t) \cdot f(t) \, dt = g(x)$

[39]

$n > \mathrm{Re}(\lambda+1) > 0, \quad \mathrm{Re}(\nu) \gtreqqless -1/2, \quad D_x^r \, g(a) = 0, \quad 0 \leqq r < n, \quad a > 0$

$f(x) = (2x)^{\nu+1} \, I_{x^2,a}^{-\nu-1} \, I_{x,a}^{\nu-\lambda} \, g(x)$

27. $\underline{\text{F.P.}} \displaystyle\int_a^x \left(x^2-t^2\right)^{\lambda/2} \, P_\nu^{-\lambda}(x/t) \, f(t) \, dt = g(x)$

[40]

$a > 0$

$f(x) = \underline{\text{F.P.}} \, (2x)^{\nu+1} \, I_{x^2,a}^{-\nu-1} \, I_{x,a}^{\nu-\lambda} \, g(x)$

5. Gegenbauer and Jacobi Polynomials

1. $$\int_0^x \left(x^2-t^2\right)^{\lambda-1/2} C_n^\lambda(x/t)\, f(t)\, dt = g(x)$$

    $$m > \text{Re}(\lambda+1/2) > 0, \quad D_x^r\, g(0) = 0, \quad 0 \leqq r < m$$

    $$f(x) = \frac{n!\, 2^{2\lambda}\Gamma(\lambda)}{\Gamma(n+2\lambda)\sqrt{\pi}}\, x^{n+1}\, I_{x^2}^{-n/2-1/2-\lambda}\, x^{1-n}\, I_{x^2}^{n/2}\, x^{-1}\, g(x)$$

2. $$\int_0^x \left(x^2-t^2\right)^{\lambda-1/2} C_n^\lambda(t/x)\, f(t)\, dt = g(x)$$

    $$m > \text{Re}(\lambda+1/2) > 0, \quad D_x^r\, g(0) = 0, \quad 0 \leqq r < m$$

    $$f(x) = \frac{n!\, 2^{2\lambda}\Gamma(\lambda)}{\Gamma(n+2\lambda)\sqrt{\pi}}\, x\, I_{x^2}^{-n/2-1/2-\lambda}\, x^{n+2\lambda}\, I_{x^2}^{n/2}\, x^{-n-2\lambda}\, g(x)$$

3. $$\int_x^b \left(t^2-x^2\right)^{\lambda-1/2} C_n^\lambda(x/t)\, f(t)\, dt = g(x)$$

    $$m > \text{Re}(\lambda+1/2) > 0, \quad D_x^r\, g(b) = 0, \quad 0 \leqq r < m, \quad x > 0, \quad b \leqq \infty$$

    $$f(x) = \frac{n!\, 2^{2\lambda}\Gamma(\lambda)}{\Gamma(n+2\lambda)\sqrt{\pi}}\, x^{n+1}\, K_{x^2}^{-n/2-1/2-\lambda}\, x^{1-n}\, K_{x^2}^{n/2}\, x^{-1}\, g(x)$$

4. $$\int_x^b \left(t^2-x^2\right)^{\lambda-1/2} C_n^\lambda(t/x)\, f(t)\, dt = g(x)$$

    $$m > \text{Re}(\lambda+1/2) > 0, \quad D_x^r\, g(b) = 0, \quad 0 \leqq r < m, \quad x > 0, \quad b \leqq \infty$$

    $$f(x) = \frac{n!\, 2^{2\lambda}\Gamma(\lambda)}{\Gamma(n+2\lambda)\sqrt{\pi}}\, x\, K_{x^2}^{-n/2-1/2-\lambda}\, x^{n+2\lambda}\, K_{x^2}^{n/2}\, x^{-n-2\lambda}\, g(x)$$

5. $\displaystyle\int_0^x (x-t)^{\lambda-1/2} \, C_n^\lambda(2x/t-1) \, f(t) \, dt = g(x)$

   $m > \text{Re}(\lambda+1/2) > 0, \quad D_x^r(0) = 0, \quad 0 \leq r < m$

$$f(x) = \frac{n! \, 2^{2\lambda-1} \Gamma(\lambda)}{\Gamma(n+2\lambda)\sqrt{\Gamma}} \, x^n \, I^{-n-2\lambda} \, x^{-n} \, I^{n+\lambda-1/2} \, g(x)$$

6. $\displaystyle\int_0^x (x-t)^{\lambda-1/2} \, C_n^\lambda(2t/x-1) \, f(t) \, dt = g(x)$

   $m > \text{Re}(\lambda+1/2) > 0, \quad D_x^r \, g(0) = 0, \quad 0 \leq r < m$

$$f(x) = \frac{n! \, 2^{2\lambda-1} \Gamma(\lambda)}{\Gamma(n+2\lambda)\sqrt{\pi}} \, x^{\lambda-1/2} \, I^{-n-2\lambda} \, x^{n+\lambda+1/2} \, I^{n+\lambda-1/2} \, x^{-n-2\lambda} \, g(x)$$

7. $\displaystyle\int_x^b (t-x)^{\lambda-1/2} \, C_n^\lambda(2x/t-1) \, f(t) \, dt = g(x)$

   $m > \text{Re}(\lambda+1/2) > 0, \quad D_x^r \, g(b) = 0, \quad 0 \leq r < m, \quad x > 0, \quad b \leq \infty$

$$f(x) = \frac{n! \, 2^{2\lambda-1} \Gamma(\lambda)}{\Gamma(n+2\lambda)\sqrt{\pi}} \, x^n \, K^{-n-2\lambda} \, x^{-n} \, K^{n+\lambda-1/2} \, g(x)$$

8. $\displaystyle\int_x^b (t-x)^{\lambda-1/2} \, C_n^\lambda(2t/x-1) \, f(t) \, dt = g(x)$

   $m > \text{Re}(\lambda+1/2) > 0, \quad D_x^r \, g(b) = 0, \quad 0 \leq r < m, \quad x > 0, \quad b \leq \infty$

$$f(x) = \frac{n! \, 2^{2\lambda-1} \Gamma(\lambda)}{\Gamma(n+2\lambda)\sqrt{\pi}} \, x^{\lambda-1/2} \, K^{-n-2\lambda} \, x^{n+\lambda+1/2} \, K^{n+\lambda-1/2} \, x^{-n-2\lambda} \, g(x)$$

9. $$\int_0^x (x-t)^\alpha \, P_n^{(\alpha,\beta)}(2x/t-1) \, f(t) \, dt = g(x)$$

$m > \text{Re}(\alpha+1) > 0, \quad \text{Re}(\beta) > -1, \quad D_x^r \, g(0) = 0, \quad 0 \overset{\leq}{=} r < m$

$$f(x) = \frac{n!}{\Gamma(\alpha+n+1)} \, x^n \, I^{-n-\alpha-\beta-1} \, x^{-n} \, I^{n+\beta} \, g(x)$$

10. $$\int_0^x (x-t)^\alpha \, P_n^{(\alpha,\beta)}(2t/x-1) \, f(t) \, dt = g(x)$$

$m > \text{Re}(\alpha+1) > 0, \quad \text{Re}(\beta) > -1, \quad D_x^r \, g(0) = 0, \quad 0 \overset{\leq}{=} r < m$

$$f(x) = \frac{n!}{\Gamma(\alpha+n+1)} \, x^\beta \, I^{-n-\alpha-\beta-1} \, x^{n+\alpha+1} \, I^{n+\beta} \, x^{-n-\alpha-\beta-1} \, g(x)$$

11. $$\int_x^b (t-x)^\alpha \, P_n^{(\alpha,\beta)}(2x/t-1) \, f(t) \, dt = g(x)$$

$m > \text{Re}(\alpha+1) > 0, \quad \text{Re}(\beta) > -1, \quad D_x^r \, g(b) = 0, \quad 0 \overset{\leq}{=} r < m, \quad x > 0,$
$$b \overset{\leq}{=} \infty$$

$$f(x) = \frac{n!}{\Gamma(\alpha+n+1)} \, x^n \, K^{-n-\alpha-\beta-1} \, x^{-n} \, K^{n+\beta} \, g(x)$$

12. $$\int_x^b (t-x)^\alpha \, P_n^{(\alpha,\beta)}(2t/x-1) \, f(t) \, dt = g(x)$$

$m > \text{Re}(\alpha+1) > 0, \quad \text{Re}(\beta) > -1, \quad D_x^r \, g(b) = 0, \quad 0 \overset{\leq}{=} r < m, \quad x > 0,$
$$b \overset{\leq}{=} \infty$$

$$f(x) = \frac{n!}{\Gamma(\alpha+n+1)} \, x^\beta \, K^{-n-\alpha-\beta-1} \, x^{n+\alpha+1} \, K^{n+\beta} \, x^{-n-\alpha-\beta-1} \, g(x)$$

6. Laguerre and Hermite Polynomials

1. $\displaystyle\int_0^x L_n(x-t)\, f(t)\, dt = g(x)$

   $g(0) = D_x\, g(0) = 0$

   $f(x) = \displaystyle\int_0^x L_{n+1}(t-x)\, e^x\, D_t^2\left[e^{-t}\, g(t)\right]\, dt$

2. $\displaystyle\int_0^x (x-t)^\alpha\, L_n^{(\alpha)}\left(k(x-t)\right)\, f(t)\, dt = g(x)$

   $m > \mathrm{Re}(\alpha) > -1, \quad D_x^r\, g(0) = 0, \quad 0 \overset{<}{=} r < m$

   $f(x) = \dfrac{n!}{\Gamma(n+\alpha+1)}\, I^{-\alpha-n-1}\, e^{kx}\, I^n\, e^{-kx}\, g(x)$

3. $\displaystyle\int_x^\beta (t-x)^\alpha\, L_n^{(\alpha)}\left(k(x-t)\right)\, f(t)\, dt = g(x)$

   $m > \mathrm{Re}(\alpha) > -1, \quad D_x^r\, g(\beta) = 0, \quad 0 \overset{<}{=} r \overset{<}{=} m, \quad x > 0, \quad \beta \overset{<}{=} \infty$

   $f(x) = \dfrac{n!}{\Gamma(n+\alpha+1)}\, K^{-\alpha-n-1}\, e^{kx}\, I^n\, e^{-kx}\, g(x)$

4. $\displaystyle\int_0^x (x-t)^m\, c_n\left(m;(x-t)\right)\, f(t)\, dt = g(x)$

   $D_x^r\, g(0) = 0, \quad 0 \overset{<}{=} r < m = n + 1$

   $f(x) = (m!)^{-1}\, D_x^{m+1}\, e^x\, I^n\, e^{-x}\, g(x)$

5. $\displaystyle\int_0^x (x-t)^{-1/2} H_{2n}\left((x-t)^{1/2}\right) f(t)\ dt = g(x)$

$g(0) = 0$

$f(x) = \dfrac{(-1)^n n!}{(2n)!\sqrt{\pi}}\ I^{-n-1/2}\ e^x\ I^n\ e^{-x}\ g(x)$

6. $\displaystyle\int_0^x H_{2n+1}\left((x-t)^{1/2}\right) f(t)\ dt = g(x)$

$g(0) = D_x\ g(0) = 0$

$f(x) = \dfrac{(-1)^n n!}{(2n+1)\sqrt{\pi}}\ I^{-n-3/2}\ e^x\ I^n\ e^{-x}\ g(x)$

7. $\displaystyle\int_x^\beta (t-x)^{-1/2} H_{2n}\left((x-t)^{1/2}\right) f(t)\ dt = g(x)$

$g(\beta) = 0,\quad x > 0,\quad \beta \overset{<}{=} \infty$

$f(x) = \dfrac{(-1)^n n!}{(2n)!\sqrt{\pi}}\ K^{-n-1/2}\ e^x\ K^n\ e^{-x}\ g(x)$

8. $\displaystyle\int_x^\beta H_{2n+1}\left((x-t)^{1/2}\right) f(t)\ dt = g(x)$

$g(\beta) = g'(\beta) = 0,\quad x > 0,\quad \beta \overset{<}{=} \infty$

$f(x) = \dfrac{(-1)^n n!}{(2n+1)!\sqrt{\pi}}\ K^{-n-3/2}\ e^x\ K^n\ e^{-x}\ g(x)$

7.   Bessel Functions

1. $\int_0^x J_0\big(a(x-t)\big)\; f(t)\; dt = g(x)$

$g(0) = D_x\, g(0) = 0$

$f(x) = \int_0^x J_0\big(a(x-t)\big)\; \left[D_t^2 + a^2\right] g(t)\; dt$

2. $\int_0^x I_0\big(a(x-t)\big)\; f(t)\; dt = g(x)$

$g(0) = D_x\, g(0) = 0$

$f(x) = \int_0^x I_0\big(a(x-t)\big)\; \left[D_t^2 - a^2\right] g(t)\; dt$

3. $\int_0^x (x-t)\; J_0\big(a(x-t)\big)\; f(t)\; dt = g(x)$

$D_x^r\, g(0) = 0,\quad 0 \overset{\le}{=} r < 3$

$f(x) = \int_0^x J_0\big(a(x-t)\big)\; \left[D_t^2 + a^2\right]^2 I_t\, g(t)\; dt$

4. $\int_0^x (x-t)\; I_0\big(a(x-t)\big)\; f(t)\; dt = g(x)$

$D_x^r\, g(0) = 0,\quad 0 \overset{\le}{=} r < 3$

$f(x) = \int_0^x I_0\big(a(x-t)\big)\; \left[D_t^2 - a^2\right]^2 I_t\, g(t)\; dt$

5. $\displaystyle\int_0^x J_1\big(a(x-t)\big)\, f(t)\, dt = g(x)$

$D_x^r\, g(0) = 0, \quad 0 \le r < 3$

$f(x) = a^{-1}\Big[D_x^2 + a^2\Big] g(x) + a^{-1}\displaystyle\int_0^x J_0\big(a(x-t)\big)\Big[D_t^3 + aD_t\Big] g(t)\, dt$

6. $\displaystyle\int_0^x I_1\big(a(x-t)\big)\, f(t)\, dt = g(x)$

$D_x^r\, g(0) = 0, \quad 0 \le r < 3$

$f(x) = a^{-1}\Big[D_x^2 - a^2\Big] g(x) + a^{-1}\displaystyle\int_0^x I_0\big(a(x-t)\big)\Big[D_t^3 - a^2 D_t\Big] g(t)\, dt$

7. $\displaystyle\int_0^x (x-t)^{-1} J_1\big(a(x-t)\big)\, f(t)\, dt = g(x)$

$g(0) = D_x\, g(0) = 0$

$f(x) = a^{-1} D_x\, g(x) + a^{-1}\displaystyle\int_0^x J_0\big(a(x-t)\big)\Big[D_t^2 + a^2\Big] g(t)\, dt$

8. $\displaystyle\int_0^x (x-t)^{-1} I_1\big(a(x-t)\big)\, f(t)\, dt = g(x)$

$g(0) = D_x\, g(0) = 0$

$f(x) = a^{-1} D_x\, g(x) + a^{-1}\displaystyle\int_0^x I_0\big(a(x-t)\big)\Big[D_t^2 - a^2\Big] g(t)\, dt$

9. $\displaystyle\int_0^x J_n\big(a(x-t)\big) \, f(t) \, dt = g(x)$

$D_x^r \, g(0) = 0, \quad 0 \stackrel{\le}{=} r < n+1$

$\displaystyle f(x) = a^{-n} \sum_{m=0}^{[(n-1)/2]} \binom{n}{2m+1} D_x^{n-2m-1} \left(D_x^2 + a^2\right)^{m+1} g(x)$

$\displaystyle \qquad + a^{-n} \int_0^x J_0\big(a(x-t)\big) \sum_{m=0}^{[n/2]} \binom{n}{2m} D_t^{n-2m} \left(D_t^2 + a^2\right)^{m+1} g(t) \, dt$

10. $\displaystyle\int_0^x I_n\big(a(x-t)\big) \, f(t) \, dt = g(x)$

$D_x^r \, g(0) = 0, \quad 0 \stackrel{\le}{=} r < n+1$

$\displaystyle f(x) = a^{-n} \sum_{m=0}^{[(n-1)/2]} \binom{n}{2m+1} D_x^{n-2m-1} \left(D_x^2 - a^2\right)^{m+1} g(x)$

$\displaystyle \qquad + a^{-n} \int_0^x I_0\big(a(x-t)\big) \sum_{m=0}^{[n/2]} \binom{n}{2m} D_t^{n-2m} \left(D_t^2 - a^2\right)^{m+1} g(t) \, dt$

11. $\displaystyle\int_0^x (x-t)^n \, J_n\big(a(x-t)\big) \, f(t) \, dt = g(x)$

$D_x^r \, g(0) = 0, \quad 0 \stackrel{\le}{=} r < 2n+2$

$\displaystyle f(x) = \frac{2^n n!}{a^n (2n)!} \int_0^x J_0\big(a(x-t)\big) \left(D_t^2 + a^2\right)^{n+1} g(t) \, dt$

12. $\displaystyle\int_0^x (x-t)^n I_n\big(a(x-t)\big) \, f(t) \, dt = g(x)$

$D_x^r g(0) = 0, \quad 0 \leqq r < 2n+2$

$f(x) = \dfrac{2^n n!}{a^n (2n)!} \displaystyle\int_0^x I_0\big(a(x-t)\big) \left(D_t^2 - a^2\right)^{n+1} g(t) \, dt$

13. $\displaystyle\int_0^x (x-t)^{n-1/2} J_{n-1/2}\big(a(x-t)\big) \, f(t) \, dt = g(x)$

$D_x^r g(0) = 0, \quad 0 \leqq r < 2n$

$f(x) = \dfrac{\sqrt{\pi}}{(2a)^{n-1/2}(n-1)!} \left(D_x^2 + a^2\right)^n g(x)$

14. $\displaystyle\int_0^x (x-t)^{n-1/2} I_{n-1/2}\big(a(x-t)\big) \, f(t) \, dt = g(x)$

$D_x^r g(0) = 0, \quad 0 \leqq r < 2n$

$f(x) = \dfrac{\sqrt{\pi}}{(2a)^{n-1/2}(n-1)!} \left(D_x^2 - a^2\right)^n g(x)$

15. $\displaystyle\int_0^x J_\nu\big(a(x-t)\big) \, f(t) \, dt = g(x)$

$|\text{Re}(\nu)| < 1, \quad g(0) = D_x \, g(0) = 0$

$f(x) = \displaystyle\int_0^x J_{-\nu}\big(a(x-t)\big) \left(D^2 + a^2\right) g(t) \, dt$

16. $\int_0^x I_\nu(a(x-t)) \; f(t) \; dt = g(x)$

$|Re(\nu)| < 1, \quad g(0) = D_x \, g(0) = 0$

$f(x) = \int_0^x I_{-\nu}(a(x-t)) \left[D_t^2 - a^2\right] g(t) \; dt$

17. $\int_0^x J_\nu(a(x-t)) \; f(t) \; dt = g(x)$

$n > Re(\nu) > -1, \quad D_x^r \, g(0) = 0, \quad 0 \leq r < n+2$

$f(x) = a^{-n}(n-\nu) \int_0^x (x-t)^{-1} \, J_{n-\nu}(a(x-t)) \sum_{m=0}^{[(n-1)/2]} \binom{n}{2m+1} D_t^{n-2m-1} \left[D_t^2 + a^2\right]^{m+1} g(t) \; dt$

$+ \; a^{-n} \int_0^x J_{n-\nu}(a(x-t)) \sum_{m=0}^{[n/2]} \binom{n}{2m} D_t^{n-2m} \left[D_t^2 + a^2\right]^{m+1} g(t) \; dt$

18. $\int_0^x I_\nu(a(x-t)) \; f(t) \; dt = g(x)$

$n > Re(\nu) > -1, \quad D_x^r \, g(0) = 0, \quad 0 \leq r < n+2$

$f(x) = a^{-n}(n-\nu) \int_0^x (x-t)^{-1} \, I_{n-\nu}(a(x-t)) \sum_{m=0}^{[(n-1)/2]} \binom{n}{2m+1} D_t^{n-2m-1} \left[D_t^2 - a^2\right]^{m+1} g(t) \; dt$

$+ \; a^{-n} \int_0^x I_{n-\nu}(a(x-t)) \sum_{m=0}^{[n/2]} \binom{n}{2m} D_t^{n-2m} \left[D_t^2 - a^2\right]^{m+1} g(t) \; dt$

19. $\displaystyle\int_0^x (x-t)^\nu J_\nu\big(a(x-t)\big) f(t)\, dt = g(x)$

$n > \mathrm{Re}(\nu+1/2) > 0, \quad D_x^r g(0) = 0, \quad 0 \stackrel{\le}{=} r < 2n$

$f(x) = \dfrac{\pi(2a)^{1-n}}{\Gamma(\nu+1/2)\Gamma(n-\nu-1/2)} \displaystyle\int_0^x (x-t)^{n-\nu-1} J_{n-\nu-1}\big(a(x-t)\big) \left[D_t^2+a^2\right]^n g(t)\, dt$

20. $\displaystyle\int_0^x (x-t)^\nu I_\nu\big(a(x-t)\big) f(t)\, dt = g(x)$

$n > \mathrm{Re}(\nu+1/2) > 0, \quad D_x^r g(0) = 0, \quad 0 \stackrel{\le}{=} r < 2n$

$f(x) = \dfrac{\pi(2a)^{1-n}}{\Gamma(\nu+1/2)\Gamma(n-\nu-1/2)} \displaystyle\int_0^x (x-t)^{n-\nu-1} I_{n-\nu-1}\big(a(x-t)\big) \left[D_t^2-a^2\right]^n g(t)\, dt$

21. $\displaystyle\int_0^x (x-t)^{\nu+1} J_\nu\big(a(x-t)\big) f(t)\, dt = g(x)$

$n > \mathrm{Re}(\nu+1/2) > -1/2, \quad D_x^r g(0) = 0, \quad 0 \stackrel{\le}{=} r < 2n+1$

$f(x) = \dfrac{\pi(2a)^{1-n}}{2\Gamma(\nu+3/2)\Gamma(n-\nu-1/2)} \displaystyle\int_0^x (x-t)^{n-\nu-1} J_{n-\nu-1}\big(a(x-t)\big) \left[D_t^2+a^2\right]^{n+1} g(t)\, dt$

22. $\displaystyle\int_0^x (x-t)^{\nu+1} I_\nu\big(a(x-t)\big) f(t)\, dt = g(x)$

$n > \mathrm{Re}(\nu+1/2) > -1/2, \quad D_x^r g(0) = 0, \quad 0 \stackrel{\le}{=} r < 2n+1$

$f(x) = \dfrac{\pi(2a)^{1-n}}{2\Gamma(\nu+3/2)\Gamma(n-\nu-1/2)} \displaystyle\int_0^x (x-t)^{n-\nu-1} I_{n-\nu-1}\big(a(x-t)\big) \left[D_t^2+a^2\right]^{n+1} g(t)\, dt$

23. $\displaystyle\int_0^x J_0\left(a(x-t)^{1/2}\right) f(t)\ dt = g(x)$

$g(0) = D_x\ g(0) = 0$

$f(x) = \displaystyle\int_0^x I_0\left(a(x-t)^{1/2}\right) D_t^2\ g(t)\ dt$

24. $\displaystyle\int_0^x I_0\left(a(x-t)^{1/2}\right) f(t)\ dt = g(x)$

$g(0) = D_x\ g(0) = 0$

$f(x) = \displaystyle\int_0^x J_0\left(a(x-t)^{1/2}\right) D_t^2\ g(t)\ dt$

25. $\displaystyle\int_0^x (x-t)^{n/2}\ J_n\left(a(x-t)^{1/2}\right) f(t)\ dt = g(x)$

$D_x^r\ g(0) = 0,\quad 0 \overset{\le}{=} r < n+2$

$f(x) = (2/a)^n \displaystyle\int_0^x I_0\left(a(x-t)^{1/2}\right) D_t^{n+2}\ g(t)\ dt$

26. $\displaystyle\int_0^x (x-t)^{n/2}\ I_n\left(a(x-t)^{1/2}\right) f(t)\ dt = g(x)$

$D_x^r\ g(0) = 0,\quad 0 \overset{\le}{=} r < n+2$

$f(x) = (2/a)^n \displaystyle\int_0^x J_0\left(a(x-t)^{1/2}\right) D_t^{n+2}\ g(t)\ dt$

27. $\int_0^x (x-t)^{\nu/2} J_\nu \left(a(x-t)^{1/2} \right) f(t) \; dt = g(x)$

       $n > \text{Re}(\nu) > -1, \quad D_x^r g(0) = 0, \quad 0 \leqq r < n+1$

$f(x) = (2/a)^{n-1} \int_0^x (x-t)^{(n-\nu-1)/2} I_{n-\nu-1} \left(a(x-t)^{1/2} \right) D_t^{n+1} g(t) \; dt$

28. $\int_0^x (x-t)^{\nu/2} I_\nu \left(a(x-t)^{1/2} \right) f(t) \; dt = g(x)$

       $n > \text{Re}(\nu) > -1, \quad D_x^r g(0) = 0, \quad 0 \leqq r < n+1$

$f(x) = (2/a)^{n-1} \int_0^x (x-t)^{(n-\nu-1)/2} I_{n-\nu-1} \left(a(x-t)^{1/2} \right) D_t^{n+1} g(t) \; dt$

29. $\int_0^x \left(x^2-t^2 \right)^{\nu/2} J_\nu \left(a \left(x^2-t^2 \right)^{1/2} \right) f(t) \; dt = g(x)$

[12] $-1 < \nu < 0$

$f(x) = a \, D_x \int_0^x \left(x^2-t^2 \right)^{-(\nu+1)/2} I_{-(\nu+1)} \left(a \left(x^2-t^2 \right)^{1/2} \right) t \; g(t) \; dt$

30. $\int_0^x \left(x^2-t^2 \right)^{\nu/2} I_\nu \left(a \left(x^2-t^2 \right)^{1/2} \right) f(t) \; dt = g(x)$

       $-1 < \nu < 0$

$f(x) = a \, D_x \int_0^x \left(x^2-t^2 \right)^{-(\nu+1)/2} J_{-(\nu+1)} \left(a \left(x^2-t^2 \right)^{1/2} \right) t \; g(t) \; dt$

31. $\int_x^\infty \left(t^2-x^2\right)^{\nu/2} J_\nu\left(a\left(t^2-x^2\right)^{1/2}\right) f(t) \, dt = g(x)$

[12]

$-1 < \nu < 0$

$f(x) = -a \, D_x \int_x^\infty \left(t^2-x^2\right)^{-(\nu+1)/2} I_{-(\nu+1)}\left(a\left(t^2-x^2\right)^{1/2}\right) t \, g(t) \, dt$

32. $\int_x^\infty \left(t^2-x^2\right)^{\nu/2} I_\nu\left(a\left(t^2-x^2\right)^{1/2}\right) f(t) \, dt = g(x)$

$-1 < \nu < 0$

$f(x) = -a \, D_x \int_x^\infty \left(t^2-x^2\right)^{-(\nu+1)/2} J_{-(\nu+1)}\left(a\left(t^2-x^2\right)^{1/2}\right) t \, g(t) \, dt$

33. $\int_0^x (x-t)^{\nu-1/2} I_{\nu-1/2}\left((x-t)/2\right) f(t) \, dt = g(x)$

[82]

$\mathrm{Re}(\nu) > 0$

$f(x) = \dfrac{\sqrt{\pi}}{2^{2\nu-1}\Gamma(\nu)} e^{x/2} I^{-\nu} e^{-x} I^{-\nu} e^{x/2} g(x)$

34. $\int_0^x t^{1+2\alpha} (t/x)^{2\eta}\left(x^2-t^2\right)^{(-\alpha-1)/2} J_{-\alpha-1}\left(k\left(x^2-t^2\right)^{1/2}\right) f(t) \, dt = g(x)$

[68]

$-m < \alpha < 0$

$f(x) = 2^m k^{2-m} x^{1-2\eta-2\alpha} \left(\dfrac{1}{2x}D_x\right)^m \left[x \int_0^x t^{1+2\eta}\left(x^2-t^2\right)^{(\alpha+m-1)/2} I_{\alpha+m-1}\left(k\left(x^2-t^2\right)^{1/2}\right) g(t) dt\right]$

35. $\displaystyle\int_0^x t^{1+2a}(t/x)^{2\eta}\left(x^2-t^2\right)^{(-\alpha-1)/2} I_{-\alpha-1}\left[k\left(x^2-t^2\right)^{1/2}\right] f(t)\ dt = g(x)$

[68]

$-m < \alpha < 0$

$\displaystyle f(x) = 2^m k^{2-m} x^{1-2\eta-2\alpha}\left(\frac{1}{2x}D_x\right)^m\left[x\int_0^x t^{1+2\eta}\left(x^2-t^2\right)^{(\alpha+m-1)/2}\right.$

$\left. J_{\alpha+m-1}\left[k\left(x^2-t^2\right)^{1/2}\right]g(t)dt\right]$

36. $\displaystyle\int_x^\infty t(x/t)^{2\eta}\left(t^2-x^2\right)^{(-\alpha-1)/2} J_{-\alpha-1}\left[k\left(t^2-x^2\right)^{1/2}\right] f(t)\ dt = g(x)$

[68]

$-m < \alpha < 0$

$\displaystyle f(x) = 2^m k^{2-m} x^{2\eta-1}\left(\frac{-1}{2x}D_x\right)^m\left[x\int_x^\infty t^{1-2\eta}\left(t^2-x^2\right)^{(\alpha+m-1)/2}\right.$

$\left. I_{\alpha+m-1}\left[k\left(t^2-x^2\right)^{1/2}\right]g(t)\ dt\right]$

37. $\displaystyle\int_x^\infty t(x/t)^{2\eta}\left(t^2-x^2\right)^{(-\alpha-1)/2} I_{-\alpha-1}\left[k\left(t^2-x^2\right)^{1/2}\right] f(t)\ dt = g(x)$

[68]

$-m < \alpha < 0$

$\displaystyle f(x) = 2^m k^{2-m} x^{2\eta-1}\left(\frac{-1}{2x}D_x\right)^m\left[x\int_x^\infty t^{1-2\eta}\left(t^2-x^2\right)^{(\alpha+m-1)/2}\right.$

$\left. J_{\alpha+m-1}\left[k\left(t^2-x^2\right)^{1/2}\right]g(t)dt\right]$

38. $\displaystyle\int_0^x k_{2n+2}\left(a(x-t)\right) f(t)\ dt = g(x)$

$g(0) = D_x\, g(0) = 0$

$\displaystyle f(x) = \frac{(-1)^n}{2a}\left(D_x+a\right)^{n+2} e^{ax} I^n e^{-ax} g(x)$

39. $\displaystyle\int_0^x \text{ber}\big(a(x-t)\big)\ f(t)\ dt = g(x)$

$D_x^r\ g(0) = 0, \quad 0 \leq r < 4$

$f(x) = \dfrac{2}{a^2} \displaystyle\int_0^x \text{bei}\big(a(x-t)\big)\left(D_t^4 + a^4\right) g(t)\ dt$

40. $\displaystyle\int_0^x \text{bei}\big(a(x-t)\big)\ f(t)\ dt = g(x)$

$D_x^r\ g(0) = 0, \quad 0 \leq r < 4$

$f(x) \quad \dfrac{2}{a^2} \displaystyle\int_0^x \text{ber}\big(a(x-t)\big)\left(D_t^4 + a^4\right) g(t)\ dt$

41. $\displaystyle\int_x^\infty (t-x)^{(\nu-1)/2}\ K_\nu\big(2(t-x)^{1/2}\big)\ f(t)\ dt = g(x)$

[141]

$\text{Re}(\nu) > 1/4$

$f(x) = \dfrac{1}{\Gamma(\nu+1)\Gamma(2\nu)} \displaystyle\int_0^\infty t^{-2}(1+t)\ {}_1F_2(\nu+3/2;\ 2\nu+1,\ \nu+1;\ -x/t)\ g(t)\ dt$

42. $\displaystyle\int_x^\infty (t-x)^{(\nu-1)/2}\ Y_\nu\big(2(t-x)^{1/2}\big)\ f(t)\ dt = g(x)$

[141]

$\text{Re}(\nu) > 1/2$

$f(x) = \dfrac{-1}{\Gamma(\nu+1/2)} \displaystyle\int_0^\infty t^{-3/2}(1+t)\ G_{13}^{21}\left(\dfrac{x}{t}\ \middle|\ \begin{matrix} -\nu-1/2 \\ 0,\ -\nu,\ 2\nu \end{matrix}\right) t(t)\ dt$

43. $\displaystyle\int_0^x I_0\left(a\left(t(x-t)\right)^{1/2}\right) f(t)\ dt = g(x)$

[69]

$\qquad g(0) = D_x\, g(0) = 0$

$f(x) = x^{-1} \displaystyle\int_0^x J_0\left(a\left(x(x-t)\right)^{1/2}\right)\left(tD_t^2 + D_t\right) g(t)\ dt$

8. Confluent Hypergeometric Functions of One and More Variables

1. $\displaystyle\int_0^x e^{\lambda(x-t)} \, \text{Erf}\left[\left(\lambda(x-t)\right)^{1/2}\right] f(t) \, dt = g(x)$

   $g(0) = D_x \, g(0) = 0$

   $f(x) = 2(\lambda\pi)^{-1/2} \, I_x^{-1/2} \, e^{\lambda x} \, D_x \, e^{-\lambda x} \, g(x)$

2. $\displaystyle\int_x^\beta e^{\lambda(t-x)} \, \text{Erf}\left[\left(\lambda(t-x)\right)^{1/2}\right] f(t) \, dt = g(x)$

   $g(\beta) = D_x \, g(\beta) = 0, \quad x > 0, \quad \beta \leq \infty$

   $f(x) = -2(\lambda\pi)^{-1/2} \, K_{x,\beta}^{-1/2} \, e^{-\lambda x} \, D_x \, e^{\lambda x} \, g(x)$

3. $\displaystyle\int_0^x \text{Erfi}\left[\left(\lambda(x-t)\right)^{1/2}\right] f(t) \, dt = g(x)$

   $g(0) = D_x \, g(0) = 0$

   $f(x) = 2(\lambda\pi)^{-1/2} \, e^{\lambda x} \, I_x^{-1/2} \, e^{-\lambda x} \, D_x \, g(x)$

4. $\displaystyle\int_x^\beta \text{Erfi}\left[\left(\lambda(x-t)\right)^{1/2}\right] f(t) \, dt = g(x)$

   $g(\beta) = D_x \, g(\beta) = 0$

   $f(x) = -2(\lambda\pi)^{-1/2} \, e^{-\lambda x} \, K_{x,\beta}^{-1/2} \, e^{\lambda x} \, D_x \, g(x)$

5. $\displaystyle\int_0^x \text{Ei}(t-x)\, f(t)\, dt = g(x)$

$g(0) = D_x\, g(0) = 0$

$f(x) = -\displaystyle\int_0^x e^{t-x}\, \nu(x-t) \left(D_t^2 + D_t \right) g(t)\, dt$

6. $\displaystyle\int_0^x \overline{\text{Ei}}(x-t)\, f(t)\, dt = g(x)$

$g(0) = D_x\, g(0) = 0$

$f(x) = -\displaystyle\int_0^x e^{t-x}\, \nu(x-t) \left(D_t^2 - D_t \right) g(t)\, dt$

7. $\displaystyle\int_0^x S(x-t)\, f(t)\, dt = g(x)$

$D_x^r\, g(0) = 0, \quad 0 \leq r < 4$

$f(x) = 4 \displaystyle\int_0^x C(x-t) \left(D_t^4 + D_t^2 \right) g(t)\, dt$

8. $\displaystyle\int_0^x C(x-t)\, f(t)\, dt = g(x)$

$D_x^r\, g(0) = 0, \quad 0 \leq r < 4$

$f(x) = 4 \displaystyle\int_0^x S(x-t) \left(D_t^4 + D_t^2 \right) g(t)\, dt$

9. $\dfrac{1}{\Gamma(n/2)} \displaystyle\int_0^x \gamma(n/2,\ x-t)\ f(t)\ dt = g(x)$

[88]
$$D_x^r\ g(0) = 0, \quad 0 \overset{<}{=} r < n+2$$

$$f(x) = \frac{1}{\Gamma(n/2)} \int_0^x \gamma(n/2,\ x-t) \cdot D_t^2 \Bigl(D_t+1\Bigr)^n\ g(t)\ dt$$

10. $\displaystyle\int_0^x \gamma(a,\ x-t)\ f(t)\ dt = g(x)$

[82]
$$D_x^r(0) = 0, \quad 0 \overset{<}{=} r < a+1$$

$$f(x) = \frac{1}{\Gamma(a)}\ e^{-x}\ I_x^{-a}\ e^x\ D_x\ g(x)$$

11. $\displaystyle\int_x^\beta \gamma(a,\ t-x)\ f(t)\ dt = g(x)$

$$D_x^r\ g(\beta) = 0, \quad 0 \overset{<}{=} r < \alpha+1, \quad x > 0, \quad \beta \overset{<}{=} \infty$$

$$f(x) = \frac{1}{\Gamma(a)}\ e^x\ K_{x,\beta}^{-a}\ e^{-x}\ D_x\ g(x)$$

12. $\displaystyle\int_0^x e^{a(x-t)}\ \Gamma\bigl(\nu, a(x-t)\bigr)\ f(t)\ dt = g(x)$

$$\mathrm{Re}(\nu) > 0, \quad g(0) = D_x\ g(0) = 0$$

$$f(x) = \frac{1}{\Gamma(\nu)} \int_0^x E_\nu\Bigl(\bigl(a(x-t)\bigr)^\nu\Bigr)\Bigl(D_t^2 - aD_t\Bigr)\ g(t)\ dt$$

13. $\displaystyle\int_0^x \Big(\cos(x-t)\ \text{Si}(x-t) - \sin(x-t)\ \text{Ci}(x-t)\Big)\ f(t)\ dt = g(x)$

$D_x^r\ g(0) = 0, \quad 0 \stackrel{\leq}{=} r < 3$

$f(x) = \displaystyle\int_0^x \nu(x-t)\left[D_t^3 + D_t\right]\ g(t)\ dt$

14. $\displaystyle\int_0^x \Big(\sin(x-t)\ \text{Si}(x-t) - \cos(x-t)\ \text{Ci}(x-t)\Big)\ f(t)\ dt = g(x)$

$g(0) = D\ g(0) = 0$

$f(x) = \displaystyle\int_0^x \nu(x-t)\left[D_t^2 + 1\right]\ g(t)\ dt$

15. $\displaystyle\int_1^x \ell i(x/t)\ f(t)\ dt = g(x)$

$g(1) = D_x\ g(1) = 0$

$f(x) = -\displaystyle\int_1^x t^{-1}(x/t)\ \nu\big(\log(x/t)\big)\left[\left(tD_t\right)^2 - \left(tD_t\right)\right]\ g(t)\ dt$

16. $\displaystyle\int_1^x \ell i(t/x)\ f(t)\ dt = g(x)$

$g(1) = D_x\ g(1) = 0$

$f(x) = -\displaystyle\int_1^x t^{-1}(t/2)\ \nu\big(\log(x/t)\big)\left[\left(tD_t\right)^2 + \left(tD_t\right)\right]\ g(t)\ dt$

17. $\int_0^x (x-t)^{(\nu-1)/2} e^{-a(x-t)/2} D_{-\nu}\left[\left(2a(x-t)\right)^{1/2}\right] f(t)\ dt = g(x)$

$|\text{Re}(\nu)| < 1, \quad g(0) = 0$

$f(x) = \pi^{-1} \int_0^x (x-t)^{(-\nu-1)/2} e^{-a(x-t)/2} D_\nu\left[\left(2a(x-t)\right)^{1/2}\right] \left(D_t + a\right) g(t)\ dt$

18. $\int_0^x \left[D_{2\nu}\left(-2\left(a(x-t)^{1/2}\right)\right) - D_{2_\nu}\left(2\left(a(x-t)\right)^{1/2}\right)\right] f(t)\ dt = g(x)$

$g(0) = D_x\ g(0) = 0$

$f(x) = \frac{\Gamma(-\nu)}{2^{\nu+1/2}\pi a}\ e^{-ax}\ I_x^{-\nu-1}\ e^{2ax}\ I_x^{\nu-1/2}\ e^{-ax}\ \dot{g}(x)$

19. $\int_0^x (x-t)^{b-1}\ {}_1F_1\left(a;\ b;\ \lambda(x-t)\right) f(t)\ dt = g(x)$

[19] $n > \text{Re}(b) > 0, \quad D_x^r\ g(0) = 0, \quad 0 \overset{<}{=} r < n$

$f(x) = \frac{1}{\Gamma(b)}\ e^{\lambda x}\ I_x^{-a}\ e^{-\lambda x}\ I_x^{a-b}\ g(x)$

20. $\int_x^\beta (t-x)^{b-1}\ {}_1F_1\left(a;\ b;\ \lambda(x-t)\right) f(t)\ dt = g(x)$

[85] $n > \text{Re}(b) > 0, \quad \text{Re}(a) < \text{Re}(b), \quad D_x^r\ g(\beta) = 0, \quad 0 \overset{<}{=} r < n, \quad x > 0, \quad \beta \overset{<}{=}$

$f(x) = \frac{1}{\Gamma(b)}\ K_{x,\beta}^{a-b}\ e^{\lambda x}\ K_{x,\beta}^{-a}\ e^{-\lambda x}\ g(x)$

21. $$\int_x^\beta (t-x)^{b-1} \, {}_1F_1\left(a; \, b; \, \lambda(x-t)\right) f(t) \, dt = g(x)$$

[85]

$$n > \text{Re}(b) > 0, \quad \text{Re}(a) > 0, \quad D_x^r g(\beta) = 0, \quad 0 \stackrel{\leq}{=} r < n, \quad x > 0, \quad \beta \stackrel{\leq}{=} \infty$$

$$f(x) = \frac{1}{\Gamma(b)} \, e^{\lambda x} \, K_{x,\beta}^{-a} \, e^{-\lambda x} \, K_{x,\beta}^{a-b} \, g(x)$$

22. $$\int_0^x (x-t)^{\mu-1} \, M_{\kappa,\mu-1/2}(x-t) \, f(t) \, dt = g(x)$$

[82]

$$n > \text{Re}(2\mu) > 0, \quad \text{Re}(\kappa) < 0, \quad D_x^r g(0) = 0, \quad 0 \stackrel{\leq}{=} r < n$$

$$f(x) = \frac{1}{\Gamma(2\mu)} \, e^{x/2} \, I^{-\mu+\kappa} \, e^{-x} \, I^{-\mu-\kappa} \, e^{x/2} \, g(x)$$

23. $$\int_0^x (x-t)^{\mu-1} \, M_{\kappa,\mu-1/2}(x-t) \, f(t) \, dt = g(x)$$

[82]

$$n > \text{Re}(2\mu) > 0, \quad \text{Re}(\kappa) = \text{Re}(\mu), \quad D_x^r g(0) = 0, \quad 0 \stackrel{\leq}{=} r < n$$

$$f(x) = \frac{1}{\Gamma(2\mu)} \, e^{-x/2} \, I^{-\mu-\kappa} \, e^{x} \, I^{-\mu+\kappa} \, e^{-x/2} \, g(x)$$

24. $$\int_x^\beta (t-x)^{\mu-1} \, M_{\kappa,\mu-1/2}\left(\lambda(t-x)\right) f(t) \, dt = g(x)$$

$$n > \text{Re}(2\mu) > 0, \quad D_x^r g(\beta) = 0, \quad 0 \stackrel{\leq}{=} r < n, \quad x > 0, \quad \beta \stackrel{\leq}{=} \infty$$

$$f(x) = \frac{1}{\lambda^\mu \Gamma(2\mu)} \, e^{-\lambda x/2} \, K_{x,\beta}^{-\mu+\kappa} \, e^{\lambda x} \, K_{x,\beta}^{-\mu-\kappa} \, e^{-\lambda x/2} \, g(x)$$

25. $\displaystyle\int_0^x (x-t)^{\mu-1} W_{\mu+k,\mu-1/2}\big(\lambda(x-t)\big)\, f(t)\ dt = g(x)$

$n > \mathrm{Re}(2\mu) > 0, \quad k \text{ an integer}, \quad D_x^r\, g(0) = 0, \quad 0 \leqq r < n$

$f(x) = \dfrac{(-1)^k}{\lambda^k \Gamma(2\mu+k)}\, e^{-\lambda x/2}\, I^{-2\mu-k}\, e^{\lambda x}\, I^k\, e^{\lambda x/2}\, g(x)$

26. $\displaystyle\int_a^x (x-t)^{\gamma-1}\, \Phi_1\big(\alpha,\beta,\gamma;\lambda(x-t),1-t/x\big)\, f(t)\ dt = g(x)$

[87]

$n > \mathrm{Re}(\gamma) > 0, \quad \mathrm{Re}(\alpha) > 0, \quad D_x^r\, g(0) = 0, \quad 0 \leqq r < n$

$f(x) = \dfrac{1}{\Gamma(\gamma)}\, I_{x,a}^{\alpha-\gamma}\, e^{\lambda x}\, x^\beta\, I_{x,a}^{-\alpha}\, e^{-\lambda x}\, x^{-\beta}\, g(x)$

27. $\displaystyle\int_0^x (x-t)^{\gamma-1}\, \Phi_2\big(\beta_1,\beta_2;\gamma;\lambda_1(x-t),\lambda_2(x-t)\big)\, f(t)\ dt = g(x)$

$n > \mathrm{Re}(\gamma) > 0, \quad D_x^r\, g(0) = 0, \quad 0 \leqq r < n$

$f(x) = \dfrac{1}{\Gamma(\gamma)}\, e^{\lambda_2 x}\, I^{-\beta_2}\, e^{\left(\lambda_1-\lambda_2\right)x}\, I^{-\beta_1}\, e^{-\lambda_1 x}\, I^{\beta_1+\beta_2-\gamma}\, g(x)$

28. $\displaystyle\int_0^x (x-t)^{\gamma-1}\, \Phi_2^r\big(\beta_1,\cdots,\beta_r;\gamma;\lambda_1(x-t),\cdots,\lambda_r(x-t)\big)\, f(t)\ dt = g(x)$

$n > \mathrm{Re}(\gamma) > 0, \quad D_x^s\, g(0) = 0, \quad 0 \leqq s < n$

$f(x) = \dfrac{1}{\Gamma(\gamma)}\, e^{\lambda_r x}\, I^{-\beta_r}\, e^{\left(\lambda_{r-1}-\lambda_r\right)x} \cdots I^{-\beta_1}\, e^{-\lambda_1 x}\, I^{\beta_1+\cdots+\beta_r-\gamma}\, g(x)$

29. $\displaystyle\int_0^x (x-t)^{\gamma-1}\,\Phi_3\big(\beta,\gamma;a(x-t),b(x-t)\big)\,f(t)\,dt = g(x)$

$n > \mathrm{Re}(\gamma) > 0, \quad D_x^r\,g(0) = 0, \quad 0 \overset{\le}{=} r < n$

$f(x) = \dfrac{1}{\Gamma(\gamma)}\,I^{\beta-\gamma-1}\,e^{ax}\,I^{-\beta}\,e^{-ax}\displaystyle\int_0^x J_0\Big(2\big(b(x-t)\big)^{1/2}\Big)\,g(t)\,dt$

30. $\displaystyle\int_0^x (x-t)^{\gamma-1}\,\Psi_1\big(\alpha,\beta,\beta,\gamma;1-a,b(x-t)\big)\,f(t)\,dt = g(x)$

$n > \mathrm{Re}(\gamma) > 0, \quad D_x^r\,g(0) = 0, \quad 0 \overset{\le}{=} r < n$

$f(x) = \dfrac{a^\alpha}{\Gamma(\gamma)}\,I^{\alpha-\gamma}\,e^{bx/a}\,I^{-\alpha}\,e^{-bx/a}\,g(x)$

31. $\displaystyle\int_0^x (x-t)^{\gamma-1}\,\Psi_1\big(\alpha,\beta,\alpha,\gamma;1-a,b(x-t)\big)\,f(t)\,dt = g(x)$

$n > \mathrm{Re}(\gamma) > 0, \quad D_x^r\,g(0) = 0, \quad 0 \overset{\le}{=} r < n$

$f(x) = \dfrac{a^\beta}{\Gamma(\gamma)}\,I^{\alpha-\gamma}\,e^{bx/a}\,I^{-\beta}\,e^{b(a-1)x/a}\,I^{\beta-\alpha}\,e^{-bx}\,g(x)$

32. $\displaystyle\int_0^x (x-t)^{\gamma-1}\,\Psi_1\big(\alpha,\beta,\beta,\alpha;1-a,b(x-t)\big)\,f(t)\,dt = g(x)$

$n > \mathrm{Re}(\gamma) > 0, \quad D_x^r\,g(0) = 0, \quad 0 \overset{\le}{=} r < n$

$f(x) = \dfrac{a^\alpha}{\Gamma(\gamma)}\,e^{bx/a}\,I^{-\gamma}\,e^{-bx/a}\,g(x)$

33. $\displaystyle\int_0^x (x-t)^{\beta-1}\ \Xi_2\!\left(\mu,1-\mu;\beta;(x-t)/2x,b^2 t(x-t)/4\right)\ f(t)\ dt = g(x)$

[76] $\qquad 0 < \mathrm{Re}(\beta) < 1, \quad g(0) = 0$

$f(x) = \dfrac{x^{-1}}{\Gamma(\beta)\Gamma(1-\beta)}\displaystyle\int_0^x t^{\beta}(x-t)^{-\beta}\ \Xi_2\!\left(\mu,1-\mu;1-\beta;(t-x)/2t,b^2 x(x-t)/4\right)$

$\qquad\qquad\qquad\qquad\qquad\qquad\qquad D_t\!\left(t^{1-\beta}g(t)\right)dt$

9. Hypergeometric Functions of One and More Variables

1.

$$\frac{1}{\Gamma(c)} \int_0^x (x-t)^{c-1} \, {}_2F_1(a,b;c;1-x/t) \, f(t) \, dt = g(x)$$

[66]

$$n > \text{Re}(c) > 0, \quad D_x^r g(0) = 0, \quad 0 \leq r < n$$

(a) $\text{Re}(b) > 0$

$$f(x) = x^{-a} \, I^{-b} \, x^a \, I^{b-c} \, g(x)$$

(b) $\text{Re}(b) < \text{Re}(c)$

$$f(x) = x^{-b} \, I^{b-c} \, x^{c-a} \, I^{-b} \, x^{a+b-c} \, g(x)$$

2.

$$\frac{1}{\Gamma(c)} \int_0^x (x-t)^{c-1} \, {}_2F_1(a,b;c;1-t/x) \, f(t) \, dt = g(x)$$

[66]

$$n > \text{Re}(c) > 0, \quad D_x^r g(0) = 0, \quad 0 \leq r < n$$

(a) $\text{Re}(b) > 0$

$$f(x) = x^{a+b-c} \, I^{-b} \, x^{c-a} \, I^{b-c} \, x^{-b} \, g(x)$$

(b) $\text{Re}(b) < \text{Re}(c)$

$$f(x) = I^{b-c} \, x^a \, I^{-b} \, x^{-a} \, g(x)$$

3. $\dfrac{1}{\Gamma(c)} \displaystyle\int_x^\infty (t-x)^{c-1}\, {}_2F_1(a,b;c;1-x/t)\, f(t)\, dt = g(x)$

[67]
$\quad n > Re(c) > 0, \quad D_x^r\, g(\infty) = 0, \quad 0 \leqq r < n, \quad x \geqq 0$

(a) $Re(b) > 0$

$f(x) = x^{-a}\, K^{-b}\, x^a\, k^{b-c}\, g(x)$

(b) $Re(b) < Re(c)$

$f(x) = x^{-b}\, K^{b-c}\, x^{c-a}\, K^{-b}\, x^{a+b-c}\, g(x)$

4. $\dfrac{1}{\Gamma(c)} \displaystyle\int_x^\infty (t-x)^{c-1}\, {}_2F_1(a,b;c;1-t/x)\, f(t)\, dt = g(x)$

[67]
$\quad n > Re(c) > 0, \quad D_x^r\, g(\infty) = 0, \quad 0 \leqq r < n$

(a) $Re(b) > 0$

$f(x) = x^{a+b-c}\, K^{-b}\, x^{c-a}\, K^{b-c}\, x^{-b}\, g(x)$

(b) $Re(b) < Re(c)$

$f(x) = K^{b-c}\, x^a\, K^{-b}\, x^{-a}\, g(x)$

5. $\dfrac{1}{\Gamma(c)} \displaystyle\int_0^x (x-t)^{\gamma-1}\, F_3(\alpha,\alpha',\beta,\beta',\gamma;1-x/t,1-t/x)\ f(t)\ dt = g(x)$

[70] $\quad n > \mathrm{Re}(\gamma) > 0,\quad D_x^r\, g(0) = 0,\quad 0 \leqq r < n,\quad \lambda \gtreqless 0$

(a) $\quad \mathrm{Re}(\gamma) > \lambda + \mathrm{Re}(\alpha') \gtreqless 0,\quad \mathrm{Re}(\alpha) \lesseqgtr 0$

$$f(x) = x^{-\beta}\, I^{-\alpha}\, x^{\beta}\, I^{\alpha+\alpha'-\gamma}\, x^{\beta'}\, I^{-\alpha'}\, x^{-\beta'}\, g(x)$$

(b) $\quad \mathrm{Re}(\gamma) > \lambda + \mathrm{Re}(\alpha) \gtreqless 0,\quad \mathrm{Re}(\alpha') \lesseqgtr 0$

$$f(x) = x^{\alpha'+\beta'-\gamma}\, I^{-\alpha'}\, x^{\gamma-\alpha-\beta'}\, I^{\alpha+\alpha'-\gamma}\, x^{\gamma-\alpha'-\beta}\, I^{-\alpha}\, x^{\alpha+\beta-\gamma}\, g(x)$$

(c) $\quad \mathrm{Re}(\gamma) \gtreqless \mathrm{Re}(\alpha+\alpha') - \lambda > 0,\quad \mathrm{Re}(\alpha) \lesseqgtr 0$

$$f(x)\quad x^{-\beta}\, I^{-\alpha}\, x^{\alpha+\alpha'+\beta+\beta'-\gamma}\, I^{-\alpha'}\, x^{\gamma-\alpha-\beta'}\, I^{\alpha+\alpha'-\gamma}\, x^{-\alpha'}\, g(x)$$

(d) $\quad \mathrm{Re}(\gamma) \gtreqless \mathrm{Re}(\alpha+\alpha') - \lambda > 0,\quad \mathrm{Re}(\alpha') \lesseqgtr 0$

$$f(x) = x^{\alpha'+\beta'-\gamma}\, I^{-\alpha'}\, x^{\gamma-\beta-\beta'}\, I^{-\alpha}\, x^{\beta}\, I^{\alpha+\alpha'-\gamma}\, x^{-\alpha'}\, g(x)$$

(e) $\quad \mathrm{Re}(\alpha+\alpha') \gtreqless \mathrm{Re}(\gamma) > \gamma + \mathrm{Re}(\alpha) \gtreqless 0$

$$f(x) = x^{-\alpha}\, I^{\alpha+\alpha'-\gamma}\, x^{\beta'}\, I^{-\alpha'}\, x^{\gamma-\beta-\beta'}\, I^{-\alpha}\, x^{\alpha+\beta-\gamma}\, g(x)$$

(f) $\quad \mathrm{Re}(\alpha+\alpha') \gtreqless \mathrm{Re}(\gamma) > \gamma + \mathrm{Re}(\alpha') \gtreqless 0$

$$f(x) = x^{-\alpha}\, I^{\alpha+\alpha'-\gamma}\, x^{\gamma-\alpha'-\beta}\, I^{-\alpha}\, x^{\alpha+\alpha'+\beta+\beta'-\gamma}\, I^{-\alpha'}\, x^{-\beta'}\, g(x)$$

10. Generalized Hypergeometric Functions

1. $$\int_0^x (x-t)^{2\mu-1} \; _1F_2\left[\nu;\mu,\mu+1/2;-\left(a(x-t)/2\right)^2\right] f(t) \; dt = g(x)$$

$$n > \text{Re}(2\mu) > 0, \quad D_x^r \; g(0) = 0, \quad 0 \leq r < n$$

$$f(x) = \frac{1}{\Gamma(2\mu)\,\Gamma(2\nu-2\mu+n)} \int_0^x (x-t)^{2\nu-2\mu+n-1}$$
$$_1F_2\left[-\nu;n/2-\nu,n/2-\mu+1/2;-\left(a(x-t)/2\right)^2\right] D_t^n \; g(t) \; dt$$

2. $$\int_0^x (x-t)^{3\mu-1} \; _1F_3\left[\nu;\mu,\mu+1/3,\mu+2/3;-\left(a(x-t)/3\right)^3\right] f(t) \; dt = g(x)$$

$$n > \text{Re}(3\mu) > 0, \quad D_x^r \; g(0) = 0, \quad 0 \leq r < n$$

$$f(x) = \frac{1}{\Gamma(3\mu)\,\Gamma(3\nu-3\mu+n)} \int_0^x (x-t)^{3\nu-3\mu+n-1}$$
$$_1F_3\left[-\nu;n/3-\mu,n/3-\mu+1/3,n/3-\mu+2/3;-\left(a(x-t)/3\right)^3\right]$$
$$D_t^n \; g(t) \; dt$$

3. $$\int_0^x \; _0F_n\left[;1/n,2/n,\cdots,(n-1)/n;1;\left((x-t)/n\right)^n\right] f(t) \; dt = g(x)$$

See $E_n\left[(x-t)^n\right]$, formula 5 of section 12

4. $\displaystyle\int_x^\infty (t-x)^{-\alpha} \, {}_pF_q\left(\alpha_1,\cdots,\alpha_p;\beta_1,\cdots,\beta_q;x-t\right) f(t) \, dt = g(x)$

[123]

$\mathrm{Re}(\beta) < \mathrm{Re}(\alpha) < 1, \quad x \overset{\geq}{=} 0, \quad x^{\alpha-\beta-1} \, g(x) \text{ and } x^{-\gamma-1}(1+1/x) \, g(x) \in L(0,\infty)$

$e_j = \alpha_{p-j+1}+\beta-1, \ j=1,\cdots,p; \ f_k = \beta_{q-k+1}+\beta-1, \ k=1,\cdots,q$

$$f(x) = \frac{\displaystyle\prod_{j=1}^{p}\Gamma\left(\alpha_j\right)}{\Gamma(\beta-\alpha)\displaystyle\prod_{j=1}^{q}\Gamma\left(\beta_j\right)} \int_0^\infty t^{-\gamma-1}(1+1/2)$$

$$\cdot \, G^{q+2,1}_{p+2,q+3}\left(\frac{x}{t}\ \middle|\ \begin{matrix} \beta-\gamma-\alpha,e_1,\cdots,e_p,\beta-\alpha \\ 0,\beta-\gamma-\alpha,f_1,\cdots,f_q,\beta \end{matrix}\right) g(t) \, dt$$

11. G and H Functions

1.
[58]

$$\int_x^1 \overline{\xi}_n(t/x) \ f(t) \ dt = g(x)$$

$$m > \operatorname{Re}\left(b_n-1\right), \quad D_x^r \ g(0) = 0, \quad 0 \overset{<}{=} r < m$$

$$\overline{\xi}_n(z) = \Gamma\left(b_n+1\right) \ G_{n,n}^{0,n}\left(z \ \middle| \ \begin{matrix} 1-a_1,\cdots,1-a_n \\ c_1,\cdots,c_n \end{matrix}\right)$$

$$\xi_{n+1}^*(z) = \Gamma\left(m-b_n+1\right) \ G_{n+1,n+1}^{n+1,0}\left(z \ \middle| \ \begin{matrix} a_1-1,\cdots,a_{k-1},0 \\ -c_1,\cdots,-c_k,-m \end{matrix}\right)$$

$$f(x) = \frac{-\Gamma\left(b_n+1\right)}{\Gamma\left(m-b_n+1\right)} \int_x^1 \xi_{n+1}^*(x/t) \ t^{-1}\left(-D_t\right)^m\left(t^{m-1}g(t)\right) dt$$

2.
[58]

$$\int_x^1 R(t/x) \ f(t) \ dt = g(x)$$

$$m > 1-\Lambda, \quad D_x^r \ g(1) = 0, \quad 0 \overset{<}{=} r < m$$

$$\Lambda = \operatorname{Re}\left(\sum_{j=1}^n b_j - \sum_{j=1}^q d_j + \frac{1}{2}(q-n)\right) < -1$$

$$\sum_{j=1}^n B_j = \sum_{j=1}^q D_j, \quad \rho = \prod_{j=1}^n B_j^{-B_j} \prod_{j=1}^q D_j^{-D_j}$$

$$R(z) = H_{n,q}^{0,n}\left[\rho z \ \middle| \ \begin{matrix} \left(1-b_1,B_1\right),\cdots,\left(1-b_n,B_n\right) \\ \left(1-d_1,D_1\right),\cdots,\left(1-d_q,D_q\right) \end{matrix}\right]$$

$$T(z) = \frac{1}{\rho^2 z} H_{n+1,n+1}^{n+1,0}\left[\rho z \ \middle| \ \begin{matrix} \left(b_1,B_1\right),\cdots,\left(b_n,B_n\right),(1,1) \\ \left(d_1,D_1\right),\cdots,\left(d_q,D_q\right),(1-m,1) \end{matrix}\right]$$

$$f(x) = \int_x^1 T(x/t) \ t^{-1}\left(-D_t\right)^m\left(t^{m-1}g(t)\right) dt$$

3.
$$\int_0^x (x-t)^{\rho-1} \, H_{p,q}^{1,n}\left[x-t \,\middle|\, \begin{matrix} (a_1,A_1),\cdots,(a_p,A_p) \\ (0,1),(b_2,B_2),\cdots,(b_q,B_q) \end{matrix} \right] f(t) \, dt = g(x)$$

[125]

See Section 4.2, Example X.

4.
$$\int_x^\infty (t-x)^{-\alpha} \, H_{p,q}^{m,n}\left[t-x \,\middle|\, \begin{matrix} (\alpha_1,A_1),\cdots,(\alpha_p,A_p) \\ (\beta_1,B_1),\cdots,(\beta_q,B_q) \end{matrix} \right] f(t) \, dt = g(x)$$

[123]

$$\mathrm{Re}(\beta) < \mathrm{Re}(\alpha) < 1 + \min \mathrm{Re}\left(\beta_j/B_j\right), \quad x \overset{\geq}{=} 0,$$

$$x^{\alpha-\beta-1} \, g(x) \quad \text{and} \quad x^{-\gamma-1}(1+1/x) \, g(x) \in L(0,\infty),$$

$$a_j = 1 - \alpha_{p-j+1} - (1-\beta) A_{p-j+1}, \quad j = 1, \cdots, p,$$

$$b_k = 1 - \alpha_{q-k+1} - (1-\beta) B_{q-k+1}, \quad k = 1, \cdots, q$$

$$f(x) = \frac{1}{\Gamma(\alpha-\beta)} \int_0^\infty t^{-\gamma-1}(1+1/t)$$

$$H_{p+2,q+2}^{q-m+2,p-n+1}\left[\frac{x}{t} \,\middle|\, \begin{matrix} (\beta-\gamma-\alpha-1),(a_1,A_1),\cdots,(a_p,A_p),(\beta-\alpha,1) \\ (0,1),(\beta-\gamma-\alpha,1),(b_1,B_1),\cdots,(b_q,B_q) \end{matrix} \right] g(t) \, dt$$

12. Miscellaneous Functions

1. $$\int_0^x \nu(x-t) \, f(t) \, dt = g(x)$$

$$g(0) = D_x \, g(0) = 0$$

$$f(x) = \int_0^x \left(\psi(1) - \log(x-t) \right) D_t^2 \, g(t) \, dt$$

2. $$\int_0^x \nu(x-t,b) \, f(t) \, dt = g(x)$$

$$n > \mathrm{Re}(b), \quad D_x^r \, g(0) = 0, \quad 0 \overset{\leq}{=} r < n+1$$

$$f(x) = \int_0^x \frac{(x-t)^{n-b-1}}{\Gamma(n-b)} \left(\psi(n-b) - \log(x-t) \right) D^{n+1} \, g(t) \, dt$$

3. $$\int_0^x h_i(x-t,n) \, f(t) \, dt = g(x)$$

$$D_x^r \, g(0) = 0, \quad 0 \overset{\leq}{=} r < i$$

$$f(x) = D_x^i \, g(x) - I_x^{n-i} \, g(x)$$

4. $$\int_0^x k_i(x-t,n) \, f(t) \, dt = g(x)$$

$$D_x^r \, g(0) = 0, \quad 0 \overset{\leq}{=} r < i$$

$$f(x) = D_x^i \, g(x) + I_x^{n-i} \, g(x)$$

5.　　$\displaystyle\int_0^x E_\alpha\left((x-t)^\alpha\right) f(t)\, dt = g(x)$

$\mathrm{Re}(\alpha) > 0, \quad g(0) = 0$

$f(x) = D_x\, g(x) - I_x^{\alpha-1}\, g(x)$

6.　　$\displaystyle\int_0^x (x-t)^{\beta-1} E_{\alpha,\beta}\left((x-t)^\alpha\right) f(t)\, dt = g(x)$

$n > \mathrm{Re}(\beta) > 0, \quad \mathrm{Re}(\alpha) > 0, \quad D_x^r\, g(0) = 0, \quad 0 \overset{\le}{=} r < n$

$f(x) = I^{-\beta}\, g(x) - I^{\alpha-\beta-1}\, g(x)$

7.　　$\displaystyle\int_a^x (x-t)^{\beta-1} E_{\alpha,\beta}^\rho\left(\lambda(x-t)^\alpha\right) f(t)\, dt = g(x)$

[84]　$n > \mathrm{Re}(\gamma) > \mathrm{Re}(\beta) > 0, \quad g^{(r)}(0) = 0, \quad 0 \overset{\le}{=} r < n$

$\displaystyle f(x) = \int_a^x (x-t)^{\gamma-\beta-1} E_{\alpha,\gamma-\beta}^{-\rho}\left(\lambda(x-t)^\alpha\right) I_{x,a}^{-\gamma}\, g(t)\, dt$

8.　　$\displaystyle\int_x^b (t-x)^{\beta-1} E_{\alpha,\beta}^\rho\left(\lambda(t-x)^\alpha\right) f(t)\, dt = g(x)$

[84]　$n > \mathrm{Re}(\gamma) > \mathrm{Re}(\beta) > 0, \quad g^{(r)}(0) = 0, \quad 0 \overset{\le}{=} r < n$

$\displaystyle f(x) = \int_x^b (t-x)^{\gamma-\beta-1} E_{\alpha,\gamma-\beta}^{-\rho}\left(\lambda(t-x)^\alpha\right) K_{x,b}^{-\gamma}\, g(t)\, dt$

9. $\displaystyle\int_0^x (x-t)^{\beta-1} \phi\left(\alpha,\beta;a(x-t)^\alpha\right) f(t)\ dt = g(x)$

[137]
$$n > Re(\beta) > 0, \quad g^{(r)}(0) = 0, \quad 0 \overset{<}{=} r < 2n$$

$$f(x) = \int_0^x (x-t)^{\beta-1} \phi\left(\alpha,\beta;-a(x-t)^\alpha\right) I^{-2\beta} g(t)\ dt$$

10. $\displaystyle\int_0^x \theta_2(0|x-t)\ f(t)\ dt = g(x)$

[15]
$$g(0) = 0$$

$$f(x) = \pi^{-1} \int_0^x \theta_3(0|x-t)\ D_t\ g(t)\ dt$$

11. $\displaystyle\int_0^x \theta_3(0|x-t)\ f(t)\ dt = g(x)$

$$g(0) = 0$$

$$f(x) = \pi^{-1} \int_0^x \theta_2(0|x-t)\ D_t\ g(t)\ dt$$

12. $\displaystyle\int_0^x A_n\left(a(x-t)\right)\ f(t)\ dt = g(x)$

$$g(0) = 0$$

$$f(x) = (-1)^n\ n!\ D_x^{n+1} \left(e^{ax} I_x\right)^n e^{-nax}\ g(x)$$

13. $\displaystyle\int_0^x \left|\sin\bigl(a(x-t)\bigr)\right| f(t)\ dt = g(x)$

$D_x^r\, g(0) = 0,\quad 0 \leq r < 3$

$f(x) = a^{-1} \displaystyle\int_0^x (-1)^{[a(x-t)/\pi]}\, D_t\!\left(D_t^2 + a^2\right) g(t)\ dt$

14. $\displaystyle\int_0^x \frac{1}{2}\!\left(\sin\bigl(a(x-t)\bigr) + \left|\sin\bigl(a(x-t)\bigr)\right|\right) f(t)\ dt = g(x)$

$D_x^r\, g(0) = 0,\quad 0 \leq r < 3$

$f(x) = a^{-1} \displaystyle\int_0^x \left(1 - U(x-t-\pi/b)\right) D_t\!\left(D_t^2 + a^2\right) g(t)\ dt$

15. $\displaystyle\int_0^x b^{-1}\!\left((x-t) - (x-t-b)U(x-t-b)\right) f(t)\ dt = g(x)$

$g(0) = g'(0) = g''(0) = 0$

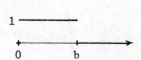

$f(x) = b \displaystyle\int_0^x \left([(x-t)/b]+1\right) g'''(t)\ dt$

16. $\displaystyle\int_0^x \left(1 - U(x-t-b)\right) f(t)\ dt = g(x)$

$g(0) = g'(0) = 0$

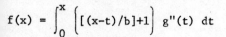

$f(x) = \displaystyle\int_0^x \left([(x-t)/b]+1\right) g''(t)\ dt$

17. $\int_0^x \left(1-2U(x-t-b) + U(x-t-2b)\right) f(t)\, dt = g(x)$

$g(0) = 0$

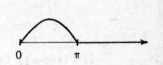

$f(x) = \sum_{0 \le n \le x/b} (n+1)\, g'(x-bn)$

18. $\int_0^x \left((x-t) -2(x-t-b)\, U(x-t-b) + (x-t-2b)\, U(x-t-2b)\right) f(t)\, dt = g(x)$

$g(0) = g'(0) = 0$

$f(x) = \sum_{0 \le n \le x/b} (n+1)\, g''(x-bn)$

19. $\int_0^x \delta^{(n)}(x-t)\, f(t)\, dt = g(x)$

$f(x) = I^n\, g(x)$

20. $\int_0^x \sin(x-t)\left(1-U(x-t-\pi)\right) f(t)\, dt = g(x)$

$g(0) = g'(0) = g''(0) = 0$

$f(x) = \frac{1}{2} \int_0^x \left[1+(-1)^{[(x-t)/\pi]}\right] \left(g'''(t)+g'(t)\right) dt$

21. $\int_0^x \cos(x-t)\left(1 - U(x-t-\pi)\right) f(t)\ dt = g(x)$

$g(0) = g'(0) = 0$

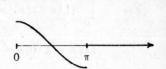

$f(x) = \frac{1}{2} \int_0^x \left(1 + (-1)^{[(x-t)/\pi]}\right) \left(g''(t) + g(t)\right)\ dt$

22. $\int_0^x \left([(x-t)/b] + 1\right) f(t)\ dt = g(x)$

$g(0) = g'(0) = 0$

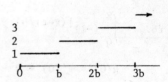

$f(x) = \int_0^x \left(1 - U(x-t-b)\right) g''(t)\ dt$

23. $\int_0^x \left(2[(x-t)/b] + 1\right) f(t)\ dt = g(x)$

$g(0) = g'(0) = 0$

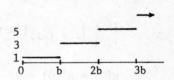

$f(x) = \int_0^x (-1)^{[(x-t)/2b]} g''(t)\ dt$

24. $\int_0^x \left(1 + (-1)^{[(x-t)/b]}\right) \left(1 - 2\left(\frac{x-t}{b} - \left|\frac{x-t}{b}\right|\right)\right) f(t)\ dt = g(x)$

$g(0) = g'(0) = g''(0) = 0$

$f(x) = \frac{b}{2} \int_0^x \left(2[(x-t)/b] + 1\right) g'''(t)\ dt$

25. $$\int_0^x \left([(x-t)/a] - [(x-t-h)/a] \right) f(t) \, dt = g(x)$$

$g(0) = g'(0) = 0, \quad h < a$

$$f(x) = \int_0^x \left([(x-t)/h] - [(x-t-a)/h] \right) g''(t) \, dt$$

26. $$\int_0^x (-1)^{[(x-t)/b]} f(t) \, dt = g(x)$$

$g(0) = g'(0) = 0$

$$f(x) = \int_0^x \frac{1}{2} \left(2[(x-t)/b] + 1 \right) g''(t) \, dt$$

27. $$\int_0^x \frac{1}{2} \left[1 + (-1)^{[(x-t)/b]} \right] f(t) \, dt = g(x)$$

$g(0) = g'(0) = 0$

$$f(x) = \int_0^x \left(1 + U(x-t-b) \right) g''(t) \, dt$$

BIBLIOGRAPHY

[1] Lothar Berg, Einführung in die Operatorenrechnung, VEB Deutscher
     Verlag der Wissenschaften, Berlin, 1965. [MR32#6164]

[2] P.L. Bharatiya, The inversion of a convolution transform whose
     kernel is a generalized Bateman's function, J. Indian Math. Soc.
     (N.S.) 28 (1964), 163-167 (1965). [MR32#6156]

[3] P.L. Bharatiya, The inversion of a convolution transform whose
     kernel is a Bessel function, Amer. Math. Monthly 72 (1965),
     393-397. [MR32#325]

[4] P.L. Bharatiya, Convolution transform with a new kernel and a
     problem of mortality of equipment, Math. Japon. 10 (1965), 45-48.
     [MR32#8054]

[5] P.L. Bharatiya, The convolution transform with the Mittag-Leffler
     function as a kernel, Amer. Math. Monthly 74 (1967), no.1 part I,
     38. [MR35#2082]

[6] B.R. Bhonsle, Inversion integrals for the Legendre transformation
     and the birth rate of a population. Ganita 17 (1966), 89-95.
     [MR39#722]

[7] B.R. Bhonsle, Inversions of some integral equations, Proc. Nat.
     Acad. Sci. India Sect. A 36 (1966), 1003-1006. [MR39#3266]

[8] T.K. Boehme, Operational calculus and the finite part of divergent
     integrals, Trans. Amer. Math. Soc. 106 (1963), 346-368.
     [MR26#1712]

[9] J. Boersma, On a function which is a special case of Meijer's
     G-function, Compositio Math. 15 (1962), 34-63. [MR24#A2683]

[10] L.S. Bosanquet, On Liouville's extension of Abel's integral
equation, Mathematika 16 (1969), 59-85. [MR40#668]

[11] B.L.J. Braaksma, Asymptotic expansions and analytic continuations
for a class of Barnes integrals, Compositio Math. 15 (1964),
239-341. [MR29#4923]

[12] J. Burlak, A pair of dual integral equations occurring in
diffraction theory, Proc. Edinburgh Math. Soc. (2) 13 (1962/3),
179-187. [MR26#5384]

[13] J. Burlak, A further note on certain integral equations of Abel
type. Proc. Edinburgh Math. Soc. (2) 14 (1964/65), 255-256.
[MR32#8091]

[14] R.G. Buschman, An inversion integral for a Legendre transformation,
Amer. Math. Monthly 69 (1962), 288-289.

[15] R.G. Buschman, An inversion integral, Proc. Amer. Math. Soc.
13 (1962), 675-677. [MR26#1704]

[16] R.G. Buschman, An inversion integral for a general Legendre
transformation, SIAM Rev. 5 (1963), 232-233. [MR27#5098]

[17] R.G. Buschman, Convolution equations with generalized Laguerre
polynomial kernels, SIAM Rev. 6 (1964), 166-167. [MR30#415]

[18] R.G. Buschman, Fractional integration, Math. Japon. 9 (1964),
99-106. [MR37#5628]

[19] R.G. Buschman, Decomposition of an integral operator by use of
Mikusiński calculus, SIAM J. Math. Anal. 3 (1972), 83-85.
[MR47#730]

[20] P.L. Butzer, Die Anwendung des Operatorenkalküls von Jan
Mikusiński auf lineare Integralgleichungen vom Faltungstypus,
Arch. Rational Mech. Anal. 2 (1958), 114-128. [MR21#2874]

[21] P.L. Butzer, Singular integral equations of Volterra type and the finite part of divergent integrals, Arch. Rational Mech. Anal. 3 (1959), 194-205. [MR21#6514]

[22] P.L. Butzer, Über den endlichen Bestandteil divergenter Integrale und singuläre Integralgleichungen von Volterra-Faltungstypus, Z. Angew. Math. Mech. 39 (1959), 355-356.

[23] N.K. Chakravarty, Review #6156 of a paper by Bharatiya, Math. Rev. 32 (1966), 1048.

[24] Ll.G. Chambers, The inversion of Legendre transforms, J. Math. Anal. Appl. 36 (1971), 172-178. [MR43#7870]

[25] Ll.G. Chambers and R.A. Sack, Expansion in spherical harmonics. V. Solution of inverse Legendre transform, J. Mathematical Phys. 11 (1970), 2197-2202. [MR43#2259]

[26] Hisachi Choda and Marie Echigo, A proof of a theorem of Widder based on the Mikusiński calculus, Amer. Math. Monthly 71 (1964), 1110-1112. [MR30#416]

[27] Ruel V. Churchill, Operational mathematics, 2nd ed., McGraw-Hill, New York, 1958. [MR21#7411]

[28] James A. Cochran, The analysis of linear integral equations, McGraw-Hill, New York, 1972.

[29] Serge Colombo, Sur les équations intégrales de Volterra à noyaux logarithmiques, C.R. Acad. Sci. Paris 235 (1952), 928-929. [MR14,562]

[30] V.A. Ditkin and A.P. Prudnikov, Integral transforms and operational calculus, (Translation) Pergamon Press, Oxford, 1965. [MR33#4609]

[31] G. Doetsch, Die Integrodifferentialgleichungen von Faltungstypus, Math. Ann. 89 (1923), 192-207.

[32] G. Doetsch, Theorie und Anwendung der Laplace-Transformation, Verlag Julius Springer, Berlin, 1937.

[33] Gustav Doetsch, Handbuch der Laplace-Transformation. Band I. Theorie der Laplace-Transformation, Verlag Birkhäuser, Basel, 1950. [MR13,230]

[34] Gustav Doetsch, Handbuch der Laplace-Transformation. Band III. Anwendung der Laplace-Transformation, 2. Abteilung, Birkhäuser Verlag, Basel, 1956. [MR18,894]

[35] A. Erdélyi, On fractional integration and its application to the theory of Hankel transforms, Quart. J. Math. Oxford Ser. 11 (1940), 293-303. [MR2,192]

[36] A. Erdélyi, On some functional transformations, Univ. e Politecnico Torino. Rend. Sem. Mat. 10 (1951), 217-234. [MR13,937]

[37] Arthur Erdélyi, Operational calculus and generalized functions, Holt, Rinehart and Winston, New York, 1962. [MR26#557]

[38] A. Erdélyi, An integral equation involving Legendre's polynomial, Amer. Math. Monthly 70 (1963), 651-652. [MR28#436]

[39] A. Erdélyi, An integral equation involving Legendre functions, J. Soc. Indust. Appl. Math. 12 (1964), 15-30. [MR29#1514]

[40] A. Erdélyi, Some integral equations involving finite parts of divergent integrals, Glasgow Math. J. 8 (1967), 50-54. [MR34#6459]

[41] A. Erdélyi, et al., Higher Transcendental Functions, I,II,III. McGraw-Hill, New York, 1953-55. [MR15,419; 16,586]

[42] A. Erdélyi, et al., Tables of Integral Transforms, I,II, McGraw-Hill, New York, 1954. [MR15,868; 16,468]

[43] Ky Fan, Exposé sur le calcul symbolique de Heaviside, Rev. Scientifique 80, nbr. 3207 (1942), 147-163. [MR7,62]

[44] István Fenyö, Über die Verallgemeinerung der Operatorenrechnung, Publ. Math. Debrecen 6 (1958), 48-59. [MR21#5865]

[45] István Fenyö, Remark on certain convolution transforms, SIAM Rev, 10 (1968), 449-450. [MR38#3698]

[46] C. Fox, A classification of kernels which possess integral transforms, Proc. Amer. Math. Soc. 7 (1956), 401-412. [MR18,36]

[47] C. Fox, The G and H functions as symmetrical Fourier kernels, Trans. Amer. Math. Soc. 98 (1961), 395-429. [MR24#A1427]

[48] E. Gesztelyi, On a generalization of the convolution, Ann. Polon. Math. 22 (1969/70), 351-358. [MR42#863]

[49] Sandy Grabiner, The use of formal power series to solve finite convolution integral equations, J. Math. Anal. Appl. 30 (1970), 415-419. [MR41#4153]

[50] G. Mustafa Habibullah, An integral equation involving Shively's polynomials, J. Natur. Sci. and Math. 10 (1970), 209-214. [MR46#7821]

[51] G. Mustafa Habibullah, Some integral equations involving confluent hypergeometric functions, Yokohama Math. J. 19 (1971), 35-43. [MR45#5691]

[52] J. Heinhold. Einige mittels Laplace-Transformation lösbare Integralgleichungen I, Math. Z. 52 (1950), 779-790. [MR12,339]

[53] Theodore P. Higgins, An inversion integral for a Gegenbauer
     transformation, J. Soc. Indust. Appl. Math. 11 (1963), 886-893.
     [MR28#2413]

[54] Theodore P. Higgins, A hypergeometric function transform, J.
     Soc. Indust. Appl. Math. 12 (1964), 601-612. [MR30#4125]

[55] Harry Hochstadt, Integral Equations, John Wiley and Sons, New
     York, 1973.

[56] A.N. Hovanskii, On a generalization of Abel's integral equation
     (Russian). Doklady Akad. Nauk SSSR (N.S.) 50 (1945), 69-70.
     [MR14,562]

[57] R.N. Jagetya, On certain integral equations involving hypergeometri
     and incomplete Gamma functions, Proc. Nat. Acad. Sci. India Sect. A
     39 (1969), 323-328. [MR44#7241]

[58] Ben C. Johnson, Integral Equations Involving Special Functions.
     Ph.D. Thesis, Oregon State Univ., June 1964. [See also Notices
     Amer. Math. Soc. 11 (1964), 742, Abstract 616-4].

[59] S.L. Kalla, On the solution of certain integral equations of
     convolution type. Acta Mexicana Ci. Tecn. 2 (1968), 85-87.
     [MR39#724]

[60] S.L. Kalla, On certain integral equations of convolution type,
     Vijnana Parishad Anusandhan Patrika 12 (1969), 73-76. [MR43#7890]

[61] Ram P. Kanwal, Linear Integral Equations, Academic Press, New
     York and London, 1971.

[62] P.R. Khandekar, On a convolution transform involving generalized
     Laguerre polynomial as its kernel, J. Math. Pures Appl. (9) 44
     (1965), 195-197. [MR31#3796]

[63] H. Kober, On fractional integrals and derivatives, Quart. J. Math. Oxford Ser. 11 (1940), 193-211. [MR2,191]

[64] Ta Li, A new class of integral transforms, Proc. Amer. Math. Soc. 11 (1960), 290-298. [MR22#3951]

[65] Ta Li, A note on integral transforms, Proc. Amer. Math. Soc. 12 (1961), 556. [MR24#A404]

[66] E.R. Love, Some integral equations involving hypergeometric functions, Proc. Edinburgh Math. Soc. (2) 15 (1967), 169-198. [MR35#7090]

[67] E.R. Love, Two more hypergeometric integral equations, Proc. Cambridge Philos. Soc. 63 (1967), 1055-1076. [MR35#6874]

[68] J.S. Lowndes, A generalization of the Erdélyi-Kober operators, Proc. Edinburgh Math. Soc. (2) 17 (1970), 139-148. [MR44#4469]

[69] A.G. Mackie, A class of integral equations, Amer. Math. Monthly 72 (1965), 956-960. [MR32#4491]

[70] O.I. Maričev, Two Volterra equations with Horn functions, Soviet Math. Dokl. 13 (1972), 703-707. (Misprints corrected from Russian Dokl. Akad. Nauk SSSR 204 (1972), 546-549). [MR46#4007]

[71] Jan Mikusiński, Operational Calculus, Pergamon Press, New York-London-Paris; Państwowe Wydawnictwo Nauke, Warsaw, 1959. [MR21#4333]

[72] J.G. Mikusiński and Cz. Ryll-Nardzewski, Sur le produit de composition, Studia Math. 12 (1951), 51-57. [MR13,231]

[73] N.W. McLachlan and Pierre Humbert, Formulaire pour le Calcul symbolique, 2d ed., Mémor. Sci. Math., no.100. Gauthier-Villars, Paris, 1950. [MR12,408]

[74] N.W. McLachlan, P. Humbert and L. Poli, Supplément au Formulaire
      pour le Calcul symbolique, Mémor. Sci. Math., no.113. Gauthier-
      Villars, Paris, 1950. [MR12,408]

[75] G. Mustafa, An integral equation involving a hypergeometric
      function, Punjab Univ. J. Math. (Lahore) 3 (1970), 35-43.
      [MR43#6679]

[76] N.E. Nørlund, Hypergeometric functions, Acta Math. 94 (1955),
      289-349. [MR17,610]

[77] Maurice Parodi, Application du calcul symbolique à la résolution
      d'une équation de Volterra dont le noyau n'appartient pas au groupe
      du cycle fermé, C.R. Acad. Sci. Paris 221 (1945), 18-19. [MR7,206]

[78] Maurice Parodi, Application du calcul symbolique à la recherche
      de solutions d'équations intégrales, Revue Sci. 85 (1947), 41.
      [MR9,40]

[79] A.S. Peters, Certain dual integral equations and Sonine's
      integrals. IMM-NYU 285, Institute of Mathematical Sciences, New
      York University, New York, 1961. 41pp. [MR42#2268]

[80] Balth. van der Pol and H. Bremmer, Operational Calculus Based
      on the Two-Sided Laplace Integral, Cambridge, at the University
      Press, 1950. [MR12,407]

[81] L. Poli, Équations intégrals et calcul symbolique, Ann. Soc.
      Sci. Bruxelles, Ser. A 55 (1935), 111-119.

[82] Tilak Raj Prabhakar, Two singular integral equations involving
      confluent hypergeometric functions, Proc. Cambridge Philos. Soc.
      66 (1969), 71-89. [MR39#3267]

[83] Tilak Raj Prabhakar, On a set of polynomials suggested by Laguerre polynomials, Pacific J. Math. 35 (1970), 213-219. [MR42#6314]

[84] Tilak Raj Prabhakar, A singular integral equation with a generalized Mittag-Leffler function in the kernel. Yokohama Math. J. 19 (1971), 7-15. [MR45#2426]

[85] Tilak Raj Prabhakar, Some integral equations with Kummer's function in the kernels, Canad. Math. Bull. 14 (1971), 391-404. [MR47#741]

[86] Tilak Raj Prabhakar, A class of integral equations with Gauss functions in the kernels, Math. Nachr. 52 (1972), 71-83. [MR47#2294]

[87] Tilak Raj Prabhakar, Hypergeometric integral equations of a general kind and fractional integration, SIAM J. Math. Anal. 3 (1972), 422-425. [MR47#742]

[88] D.O. Reudink, Convolution transforms whose inversions have the same kernel, SIAM Rev. 9 (1967), 721-725. [MR36#5622]

[89] G.E. Roberts and H. Kaufman, Table of Laplace Transforms, W.B. Saunders, Philadelphia, 1966. [MR32#8050]

[90] Fritz Rühs, Operatorenrechnung und Hadamardscher Partie finie, Math. Nachr. 30 (1965), 237-250. [MR32#8072]

[91] K.C. Rusia, An integral equation involving Jacobi polynomial, Proc. Nat. Acad. Sci. India Sect A 36 (1966), 933-936. [MR39#7374]

[92] K.C. Rusia, An integral equation involving generalized Laguerre polynomial. Math. Japon. 11 (1966), 15-18. [MR34#1800]

[93] K.C. Rusia, Some integral equations and integrals, Proc. Nat. Acad. Sci. India Sect. A 37 (1967), 67-70. [MR40#3232]

[94] K.C. Rusia, An integral equation involving confluent hypergeometric function, Math. Student 37 (1969), 55-58. [MR41#8951]

[95] K.C. Rusia, A class of integral equations, Proc. Nat. Acad. Sci. India Sect. A 39 (1969), 334-336. [MR44#755]

[96] K.C. Rusia, A class of integral equations involving confluent hypergeometric functions, Proc. Nat. Acad. Sci. India Sect. A 39 (1969), 349-354. [MR44#756]

[97] K.C. Rusia, On some integral equations involving Jacobi polynomials, Proc. Nat. Acad. Sci. India Sect. A 39 (1969), 381-388. [MR44#757]

[98] K.C. Rusia, On certain integral equations, Ganita 21 (1970), 33-37. [MR45#4086]

[99] K.C. Rusia, Application of an integral equation involving generalized Laguerre polynomial in the evaluation of certain integrals and in the problem of mortality of equipment, Proc. Nat. Acad. Sci. India Sect. A 40 (1970), 177-180. [MR46#7594]

[100] R.A. Sack, Expansion in spherical harmonics. IV. Integral form of the radial dependence, J. Mathematical Phys. 8 (1967), 1774-1784. [MR42#2258]

[101] L.A. Sahnovich, Spectral analysis of operators of the form $kf = \int_0^x f(t) \, k(x-t) \, dt$, Izv. Akad. Nauk SSSR Ser. Math. 22 (1958), 299-308. [MR20#5409]

[102] S.G. Samko, On the relation of certain integral equations of the first kind in the theory of elasticity and hydrodynamics to equations of the second kind. (Russian) Prik. Mat. Met. 31 (1967), 343-345, Transl. (J. Appl. Math. Mech. 31 (1967), 368-371). [MR43#730]

[103] K.S. Sevaria, An integral equation involving confluent hypergeometric series of two variables, Univ. Nac. Tucumán Rev. Ser. A 19 (1969), 55-57. [MR40#7754]

[104] M.T. Shah, Inversion of a convolution transform whose kernel is a Hermite polynomial (Arabic summary), Bull. College Sci. (Baghdad) 10 (1967), 39-42 (1968). [MR39#4612]

[105] C. Singh, On a convolution transform involving Whittaker's function as its kernel. Math. Japon. 13 (1968), 71-74. [MR40#4702]

[106] C. Singh, Inversion integrals for the integral transforms with Bessel functions in the kernel, Riv. Mat. Univ. Parma (2) 9 (1968), 339-341.

[107] C. Singh, Inversion of a convolution transform with Kelvin's function as a kernel, Proc. Nat. Acad. Sci. India Sect. A 39 (1969), 279-280. [MR44#734]

[108] C. Singh, Two inversion integrals, Riv. Mat. Univ. Parma (2) 11 (1970), 165-168.

[109] C. Singh, An inversion integral for a Whittaker transform, Riv. Mat. Univ. Parma (2) 11 (1970), 277-280. [MR47#743]

[110] C. Singh, An integral equation with Jacobi polynomial in the kernel, Riv. Mat. Univ. Parma (2) 11 (1970), 313-316. [MR47#744]

[111]   C. Singh, An integral equation with Jacobi polynomial in the
        kernel, J. Ravishankar Univ. 1 (1972), 9-10.

[112]   R.P. Singh, An integral equation involving generalized Legendre
        polynomials, Math. Student 35 (1967), 81-84 (1969). [MR40#6215]

[113]   P. Skibiński, Som expansion theorems for the H-function, Ann.
        Polon. Math. 23 (1970), 125-138. [MR42#2058]

[114]   C.V.L. Smith, The fractional derivative of a Laplace integral,
        Duke Math. J. 8 (1941), 47-77. [MR2,281]

[115]   I.N. Sneddon, Mixed Boundary Value Problems in Potential Theory,
        North-Holland, Amsterdam, 1966. [MR35#6853]

[116]   Ian N. Sneddon, A procedure for deriving inversion formulae
        for integral transform pairs of a general kind, Glasgow Math.
        J. 9. (1968), 67-77. [MR37#4516]

[117]   Ian N. Sneddon, The Use of Integral Transforms, McGraw-Hill,
        New York, 1971.

[118]   Kusum Soni, A Sonine transform, Duke Math. J. 37 (1970),
        431-438. [MR41#7389]

[119]   Kusum Soni, A unitary transform related to some integral
        equations, SIAM J. Math. Anal. 1 (1970), 426-436. [MR43#5271]

[120]   Kusum Soni, An integral equation with Bessel function kernel,
        Duke Math. J. 38 (1971), 175-180. [MR43#851]

[121]   R.P. Srivastav, A note on certain integral equations of Abel-type,
        Proc. Edinburgh Math. Soc. (2) 13 (1962/3), 271-272. [MR27#6096]

[122]   R.P. Srivastav, On certain integral equations of convolution
        type with Bessel-function kernels, Proc. Edinburgh Math. Soc.
        (2) 15 (1966/7), 111-116. [MR36#4300]

[123] H.M. Srivastava, A class of integral equations involving the H function as kernel, Nederl. Akad. Wetensch. Proc. Ser. A 75 = Indag. Math. 34 (1972), 212-220.

[124] H.M. Srivastava, An integral equation involving the confluent hypergeometric function of several complex variables, Applicable Anal. 5 (1976), 251-256 ; see also Notices Amer. Math. Soc. 18 (1971), 1100, Abstract 71T-B246.

[125] H.M. Srivastava and R.G. Buschman, Some convolution integral equations, Nederl. Akad. Wetensch. Proc. Ser. A 77 = Indag. Math. 36 (1974), 211-216.

[126] K.N. Srivastava, On some integral transforms, Math. Japon. 6 (1961/2), 65-72. [MR27#1785]

[127] K.N. Srivastava, A class of integral equations involving ultraspherical polynomials as kernels, Proc. Amer. Math. Soc. 14 (1963), 932-940. [MR27#6102]

[128] K.N. Srivastava, Inversion integrals involving Jacobi's polynomials, Proc. Amer. Math. Soc. 15 (1964), 635-638. [MR29#292]

[129] K.N. Srivastava, A note on a class of integral equations involving ultraspherical polynomials as kernel, Math. Japon. 9 (1964), 83-84. [MR37#1905]

[130] K.N. Srivastava, On some integral equations involving Jacobi polynomials, Math. Japon. 9 (1964), 85-88. [MR37#1906]

[131] K.N. Srivastava, Integral equations involving a confluent hypergeometric function as kernel, J. Analyse Math. 13 (1964), 391-397. [MR32#8066]

[132] K.N. Srivastava, Fractional integration and integral equations
       with polynomial kernels, J. London Math. Soc. 40 (1965), 435-440.
       [MR31#2573]

[133] K.N. Srivastava, A class of integral equations involving Jacobi
       polynomials as kernel, Proc. Nat. Acad. Sci. India Sect. A 35
       (1965), 221-226. [MR33#6327]

[134] K.N. Srivastava, On some integral transforms involving Jacobi
       functions, Ann. Polon. Math. 16 (1965), 195-199. [MR30#4124]

[135] K.N. Srivastava, A class of integral equations involving Laguerre
       polynomials as kernel, Proc. Edinburgh Math. Soc. (2) 15 (1966),
       33-36. [MR34#3241]

[136] K.N. Srivastava, On integral equations involving Whittaker's
       function, Proc. Glasgow Math. Assoc. 7 (1966), 125-127.
       [MR32#8097]

[137] T.N. Srivastava and Y.P. Singh, On Maitland's generalized Bessel
       function, Canad. Math. Bull. 11 (1968), 739-741. [MR40#654]

[138] Dietrich Suschowk, Die Umkehrung einer Klass singulärer
       Integrale, Math. Z. 69 (1958), 363-365. [MR21#274]

[139] Mikiharu Terada, On an inversion formula for an integral
       transform of convolution type, Mem. Osaka Univ. Lib. Arts Ed.
       Ser. B No. 13 (1964), 27-31. [MR31#3797]

[140] E.C. Titchmarsh, Introduction to the theory of Fourier Integrals,
       2nd ed., Oxford, at the Clarendon Press, 1948.

[141] V.K. Varma, Inversion of a class of transform with a difference
       kernel, Proc. Cambridge Philos. Soc. 65 (1969), 673-677.
       [MR38#4925]

[142]   Richard Weiss,   Product Integration for the generalized Abel
        equation,   Math. Comp. 26 (1972), 177–189.   [MR45#8050]

[143]   D.V. Widder,   The inversion of a convolution transform whose
        kernel is a Laguerre polynomial,   Amer. Math. Monthly 70 (1963),
        291–293.   [MR36#6699]

[144]   D.V. Widder,   Two convolution transforms which are inverted
        by convolutions,   Proc. Amer. Math. Soc. 14 (1963), 812–817.
        [MR27#4022]

[145]   Klaus Wiener,   Über die Lösung einiger Integralgleichungen mit
        Hadamard-Integraler,   Wiss. Z. Martin-Luther Univ. Halle-
        Wittenburg Math-Natur. Riehe 15 (1966), 645–660.   [MR35#4694b]

[146]   Klaus Wiener,   Zür Lösung der Abelschen Integralgleichung mit
        einen Hadamard-Integralen,   Wiss. Z. Martin-Luther-Univ. Halle-
        Wittenburg Math-Natur. Riehe 15 (1966), 661–666.   [MR35#4694c]

[147]   Jet Wimp,   Two integral transform pairs involving hypergeometric
        functions,   Proc. Glasgow Math. Assoc. 7 (1965), 42–44.
        [MR31#1402]

AUTHOR INDEX

SUBJECT INDEX